African Media and the Digital Public Sphere

Palgrave Macmillan Series in International Political Communication

Series editor:

Philip Seib, University of Southern California

From democratization to terrorism, economic development to conflict resolution, global political dynamics are affected by the increasing pervasiveness and influence of communication media. This series examines the participants and their tools, their strategies and their impact. It offers a mix of comparative and tightly focused analyses that bridge the various elements of communication and political science included in the field of international studies. Particular emphasis is placed on topics related to the rapidly changing communication environment that is being shaped by new technologies and new political realities. This is the evolving world of international political communication.

Books Appearing in this Series

Media and the Politics of Failure: Great Powers, Communication Strategies, and Military Defeats
By Laura Roselle

The CNN Effect in Action: How the News Media Pushed the West toward War in Kosovo
By Babak Bahador

Media Pressure on Foreign Policy: The Evolving Theoretical Framework
By Derek B. Miller

New Media and the New Middle East
Edited by Philip Seib

The African Press, Civic Cynicism, and Democracy
By Minabere Ibelema

Global Communication and Transnational Public Spheres
By Angela M. Crack

Latin America, Media, and Revolution: Communication in Modern Mesoamerica
By Juanita Darling

Japanese Public Opinion and the War on Terrorism
Edited by Robert D. Eldridge and Paul Midford

African Media and the Digital Public Sphere
Edited by Okoth Fred Mudhai, Wisdom J. Tettey, and Fackson Banda

African Media and the Digital Public Sphere

Edited by

Okoth Fred Mudhai, Wisdom J. Tettey, and Fackson Banda

palgrave
macmillan

AFRICAN MEDIA AND THE DIGITAL PUBLIC SPHERE

First published in 2009 by
PALGRAVE MACMILLAN®
in the United States—a division of St. Martin's Press LLC,
175 Fifth Avenue, New York, NY 10010.

Where this book is distributed in the UK, Europe and the rest of the world, this is by Palgrave Macmillan, a division of Macmillan Publishers Limited, registered in England, company number 785998, of Houndmills, Basingstoke, Hampshire RG21 6XS.

Palgrave Macmillan is the global academic imprint of the above companies and has companies and representatives throughout the world.

Palgrave® and Macmillan® are registered trademarks in the United States, the United Kingdom, Europe and other countries.

ISBN: 978–0–230–61486–4

Library of Congress Cataloging-in-Publication Data

African media and the digital public sphere / edited by Okoth Fred Mudhai, Wisdom J. Tettey, and Fackson Banda.
p. cm.—(Palgrave Macmillan series in international political communication)
Includes bibliographical references.
ISBN 978–0–230–61486–4
1. Information society—Africa. 2. Democracy—Africa. 3. Information technology—Social aspects—Africa. 4. Information technology—Political aspects—Africa. I. Mudhai, Okoth Fred. II. Tettey, Wisdom. III. Banda, Fackson.

HN780.Z9I5625 2009
306.2096′090511—dc22 2008043027

A catalogue record of the book is available from the British Library.

Design by Newgen Imaging Systems (P) Ltd., Chennai, India.

First edition: June 2009

Contents

Tables

Preface

This book examines, from theoretical and empirical perspectives, the claims that new information and communication technologies (ICTs) are catalysts of democratic change in Africa. Contributors do so from optimist, pragmatist-realist, and pessimist stances through analyses of various forms of evidences—including words and deeds of various political actors and organizations or institutions, from government units to political parties and party leaders to civil society organizations and minority or marginalized groups. The main focus is, therefore, on the interrelated concepts of e-participation and e-democracy. The UN defines e-participation as "participatory, inclusive, deliberative process of decision-making," that is the use of ICTs to disseminate information (e-information), to seek views and encourage discussions (e-consultation) and to take these ideas into account when making decisions that affect citizens (e-decision-making) (Misuraca 2007, ch.1 sec.1.3).[1] While Misuraca (2007, ch.1 sec.1.3) notes "there are as many interpretations of what constitutes e-democracy as there are interpretations of democracy," prolific e-democracy writer Stephen Clift perceives the concept as representing "the use of ICTs and strategies by democratic actors within political and governance processes...Democratic actors/sectors include governments, elected officials, the media, political organizations, and citizen/voters" (Misuraca 2007, ch.1 sec.1.3). It is these actors that authors of this book turn their lenses on, chapter by chapter.

The book is—to our knowledge, by the time of production—the first such academic publication contributed to by various African and Africanist scholars, based in Africa and around the world, whose research and/or practice activities focus on the relationship between new digital media, converged or otherwise, and democracy on the continent. It is a contextual examination of theoretical perspectives and empirical evidences on the potential and real impact of new media

---

[1] See reference in chapter 2. Citation before the last paragraph of the chapter, "The Potential of ICTs for Good Governance: Basic Concepts and Definitions", page 5 of 16 in G. Misuraca (2007). http://www.idrc.ca/en/ev-115660-201-1-DO_TOPIC.html (accessed December 20, 2008). For more on Clift, see: www.publicus.net

on democracy in Africa. Contributors use various approaches to democratic theory in examining—endorsing or challenging—the notion that ICTs provide opportunities for greater democracy in Africa. They examine deeply the notion that new media instigate, even if more exogenously than endogenously, changes that influence, amplify, or magnify the conduct of politics in important but often unanticipated ways. Authors explore how much of Africa's moves toward democratization could be attributed to new media—taking into account other equally, and possibly more, influential historical and contemporary socioeconomic and political factors, such as violence, that impact on politics.

The idea of convergence is vital in the conception of new media, which in this book refers to recent digital computer-based ICTs, especially the World Wide Web (the Web), e-mail, and cell phones, and their employment as adjuncts or boosters of more conventional or mainstream news and communication media such as the press, radio, and TV. Interactivity and flexibility (with less limitations on target/scope) are key ingredients of new media that are more difficult, but of course not impossible, to censor. The focus is not so much on (McLuhannisque) emphasizing technology of the medium, but on the use to which political institutions or organizations and actors put new media within specific socioeconomic and political contexts.

Some of the chapters were presented as papers at a panel of the Media, Communication, and Cultural Studies Association conference at Coventry University's Technology Centre in January 2007. Others were presented at an academic seminar organized by the SAB LTD—UNESCO Chair of Media and Democracy, Prof. Fackson Banda, held on September 9, 2006, at Rhodes University, Grahamstown, South Africa. Indeed, some—if not all—of the questions Banda posed ahead of the Rhodes seminar are of central significance for this book:

> Given Africa's positioning in relation to new technologies, can we legitimately begin to talk about the emergence of a "digital public sphere"? If the Habermasian conceptualization of the "public sphere" is largely in terms of "rationality" and "scale," who then is participating in this digital public sphere, and from whence does it get its "legitimacy"? Is there any empirical evidence to suggest that Africa is moving towards any such sphere? What forms of mediation can best advance a truly democratic digital public sphere? What models of democracy can best promote the attainment of such a sphere? Indeed, what type of digital public sphere can best advance democratic ideals? (Email message call to potential seminar presenters and participants, January 19, 2006).

In addressing these and related questions, chapter authors link digital democracy to the public sphere concept.[2] The multidisciplinary nature of contributors (including sociologists, journalists, and other media people) not only injects richness in the material but also shows the varied interests this book will attract among learners, tutors, and researchers at various levels of college and university education—especially in the areas of politics, development, and area (African) studies—as well as among democracy practitioners, donors, and observers.

OKOTH FRED MUDHAI

---

[2] The wide literature on public sphere is most recently summarized and critiqued in Chapter 1 of Matthew Hindman's *The Myth of Digital Democracy* (2008, Princeton University Press). See also Angela M. Crack's *Global Communication and Transnational Public Spheres* (2008, Palgrave Macmillan). I attempt to give an overview of African approaches in: 'Researching the impact of ICTs as change catalysts in Africa' (*Ecquid Novi: AJS* Vol. 25 No.2, 313–335, 2004).

Notes on Contributors

Khayrat Ayyad is an Assistant Professor and the Head of Public Relations Department in the College of Communication, University of Sharjah, Sharjah, United Arab Emirates. He holds a PhD from the Centre for Mass Communication Research, University of Leicester, UK.

Fackson Banda is South African Breweries (SAB) LTD-UNESCO Chair of Media and Democracy at the School of Journalism and Media Studies, Rhodes University, South Africa. Winner of the 2007 MISA Zambia Media Freedom Award and the 2008 MISA Press Freedom Award for southern Africa, Prof. Banda was previously a broadcast journalist and, until 2005, director of Panos Southern Africa.

Carol Azungi Dralega is a Community Media Research Consultant with Last Mile Consulting, Norway. She has worked as a newspaper journalist and editor in Uganda and as a lecturer and researcher with the department of Media and Communication, University of Oslo, Norway. Her PhD (2008) thesis in the same university focused on new media uses by grassroots communities, women, and the youth—especially in Uganda.

Fabian Frenzel, formerly of the Free University of Berlin (Germany), is a PhD Researcher at the Centre for Tourism and Cultural Change, Leeds Metropolitan University, UK. His research focuses on the mobilities of political tourists and social movements. He has also worked with Indymedia in Europe and in Africa.

Marie-Soleil Frère is a Research Associate at Belgium's National Fund for Scientific Research and also Professor of Journalism, specializing in the African media, at the University of Brussels. She is an associate expert for the Institut Panos Paris.

Alain Kiyindou is a Senior Lecturer in Information and Communication Sciences and researcher at the Research Center in Social Sciences, University of Strasbourg, France.

Patricia K. Litho is the Head of Social Science Professional Grouping at Cavendish University, Uganda. She is also a communications consultant. Her interests, pursued with a PhD at the University of East London, include ICTs for development and gender issues. She has worked as a broadcast journalist in Uganda.

Last Moyo is an Assistant Professor in the Department of Media, Peace and Conflict Studies at Hankuk University of Foreign Studies, South Korea. Before obtaining his PhD at the University of Wales, Aberystwyth, he taught at a university in Bulawayo, Zimbabwe.

Okoth Fred Mudhai is a Senior Lecturer in Journalism and Global Media/Communication at Coventry University, UK. He was part of IT and Civil Society Network of the IT and International Cooperation Program, U.S. Social Science Research Council (2003–2005). His PhD (the Nottingham Trent University, UK) and MA (Leeds University, UK) theses focused on ICT access-usage in Africa. He has won three media awards, two on ICTs at global level.

George Nyabuga is a Lecturer in Media and Cultural Studies at the University of Worcester, UK. He wrote his PhD on ICTs and Democracy in the Department of Politics, Coventry University, UK. He previously worked as a journalist in Kenya, South Africa, and the United States.

George Ogola is a Senior Lecturer in Journalism at the University of Central Lancashire, UK. He did his PhD at the University of Witwatersrand, South Africa. He has worked as a journalist in Kenya and South Africa.

Michael Olutayo Olatunji is a Senior Lecturer of Art Music (Performance, Composition, and Orchestration) at Obafemi Awolowo University Ile-Ife, Nigeria. His PhD (2006) focussed on Military Music in Nigeria.

Rune Ottosen is a Professor in Journalism at Oslo University College, Norway. He has published several books and articles within the field of war-and-peace journalism, digital media, and press history.

Anne Schumann is a PhD Researcher at the School of Oriental and African Studies, University of London. She holds an MSc in African Politics and a BA in African Studies and English Literature. She has worked partly with the publishing industry.

Sian Sullivan is a Lecturer in Environment and Development at Birkbeck College, University of London. She studies social movements, cultural landscapes, shamanism, ethnobiology, conservation, and the privatization of biodiversity.

Wisdom J. Tettey is a Professor in the Faculty of Communication and Culture, University of Calgary, Canada. His research expertise is in the areas of mass media and democratic change in Africa; ICTs, civic engagement and globalization; transnationalism and the African Diaspora.

Keyan G. Tomaselli is the Deputy Dean (Research) and Professor in Culture, Communication, and Media Studies, University of KwaZulu-Natal, Durban, South Africa.

Ruth E. Teer-Tomaselli is a Senior Professor in Culture, Communication, and Media Studies, University of KwaZulu-Natal, Durban, South Africa. She has also worked at senior levels in the South African Broadcasting Corporation.

Kutoma J. Wakunuma-Zojer is a Research Associate at Sheffield Hallam University, UK—investigating Technology Design and Social Action. She has also worked as a Research Fellow of the Initiativkreis Ruhrgebiet, Witten/Herdecke University, Germany. Her PhD from Coventry University, UK covered ICTs—specifically the Internet and mobile phones.

CHAPTER 1

Introduction: New Media and Democracy in Africa—A Critical Interjection

Fackson Banda, Okoth Fred Mudhai, and Wisdom J. Tettey

The advent of new media technology in Africa, in the 1990s, sparked celebratory, almost utopian bliss in its proponents. It was accompanied by the hype about the continent's possibility of "leapfrogging" some stages of development, as though the whole process of development had been rendered less problematic. A linearity of progress was assumed, almost uncritically positing new media technology as *deterministic* of social progress. Yet, after over a decade of the emergence of new information and communication technologies (ICTs), the old questions about access, inequality, power, and the quality of information available are still valid (cf. Fourie, 2001). This calls for a more critical rethink of the social and political impact of new technologies on the African polity. This book makes a significant contribution to the discourse around these questions by bringing together a collection of chapters that explore the correlation between new media technology and democracy in Africa, as well as the nature of their juxtaposition with "old" or "traditional" media. To set the context for these analyses, this introductory chapter provides a critical analysis of the conceptual and theoretical debates surrounding the new media/political engagement/democratic participation/good-governance nexus, and anchors them in the specific reality of the African situation.

Discussions of new media often disregard the unusual African terrain, which defies many of the technological innovations said to be reconfiguring the structures and processes of communication globally. Of significance are poor telecommunication networks in most parts of Africa, resulting in a low level of Internet usage (cf. Adeya and Cogburn 2001, 77–109). Urging great caution in appraising new ICTs is not to suggest that Africa's appropriation of technology is hopeless; there is anecdotal evidence, as we demonstrate later, of the emergence of online communities across Africa. Although such communities are mainly associated with institutional elites, they are still indicative of an emerging community of Africans taking advantage of some of the novel features of new technology to forge and advance their own identities and agendas.

Perhaps the value of new media in Africa lies in the fact that they have supplemented, but by no means supplanted, old media forms. As Martin Lister et al. (2003, 10) suggest, discussions about "new media" force us to acknowledge, on one hand, a rapidly changing set of formal and technological experiments and, on the other, a complex set of interactions between new technological possibilities and established media forms. As George Landow (2003, 58) suggests, "new media" technologies should be seen as existing on a continuum or spectrum rather than in any fundamental opposition to one another. This perspective allows us to contextualize "new media"—the Internet, e-mail, cellular telephony, and so on—in terms of the old questions posed of "old media," such as universal access, regulation and content. The value of new media thus lies in the extent to which they enmesh with old media to provide multimedia platforms that allow for greater democratic participation, inclusion, and expression.

Exploring the (Dis)Connections between "New" and "Old" Media

What are the features that new media offer as potential extensions of old media? Lister et al. (2004, 10–13) list, inter alia, the following: the intensity of change typical of new media; the ideological connotations of the new; and the nontechnical and inclusive nature of new media platforms. We elaborate upon these.

The Intensity of Change Associated with New Media

The term "new media" has about it an aura of change, a change that was quite rapid especially from the late 1980s. Several such changes are associated with new media. First, new media are seen as having sped up the deconstructive element of postmodernity, a contested but widely subscribed attempt to characterize deep and structural changes in societies and economies from the 1960s onward, with correlative cultural changes. For example, the perceived shift from elitist forms of democracy to participatory forms of democratic expression are associated with the less hierarchized nature of new media technology. In a sense, then, new media technology is seen as subverting old hierarchies of communication, thereby decentering the very act of communicating.

Second, new media technology is perceived as heralding the era of globalization, conceived of as both a cause and an effect of technological advancement. Lister et al. (2003) suggest that new media have been seen as a contributory element of globalization, which entails a dissolving of national states and boundaries. This perspective accords with Appadurai's view of globalization as representing a series of interlocking scapes—"ethnoscape," "technoscape," "mediascape," "ideoscape," and "financescape" (Appadurai 1996; cf. Robertson 1992). However, we must be careful here not to generalize the rupturing of national states and boundaries to Africa. While most of Africa has become increasingly interconnected with the rest of the world, it is evident that the nation-state still looms large and controls much of the geography and infrastructure of communications (Banda 2003 and 2006).

Third, new media technology has arguably heralded the "post-industrial" information age, a shift to information "industries" that many uses of new media are seen to epitomize (Lister et al. 2003). Finally, new media technology is perceived as underpinning the process of decentralization, the weakening of mechanisms of power and control from dominant centers of power, facilitated by the dispersed, boundary-transgressing, networks of new communication media. It can also mean the decentering of power within African nation-states, from urban "centres" to rural "metropoles." While this can be said of general communications, it cannot be said of the structures and processes of political communication. A majority of African countries are still locked in systems of political-communicational centralization. In summary, then, new media are seen as part of a much larger landscape of social, technological, and cultural change, what Lister et al. (2003, 11) refer to as a "new technoculture."

The Ideological Connotations of "the New"

The sense of the "new" in new media as a reference to "the most recent" also carries the ideological sense that new equals better and is glamorous and exciting. It is "the cutting edge," the "avant-garde," for forward-thinking people—derived from a modernist belief in social progress as delivered by technology (Lister et al. 2003, 11). New media appear, as they have before, with claims and hopes attached—about ideologies of progress for the West and the possibilities of the Third World "catching up" (Lister et al. 2003, 11; cf. Banda 2003; Fourie 2001).

This apparently innocent enthusiasm for the "latest" is rarely if ever ideologically neutral and cannot be dissociated from the globalizing neoliberal forms of production and distribution characteristic of the past twenty years (Lister et al. 2003, 11). Such celebratory approaches to new media technology hardly go uncontested in the African context; indeed it does not go uncontested even in the contexts of the developed countries. As Roszak (2004, 61), expanding upon his notion of "technophilia," observes:

> Such technological infatuations come and go... I think, however, that the current fascination with the computer and its principal product, information, deserves a more critical response... the computer does so ingeniously mimic human intelligence that it may significantly shake our confidence in the uses of the mind. And it is the mind that must think about all things, including the computer.

Nontechnical and Inclusive Aspect of New Media

Ascribed to "new media" is the quality of inclusiveness. It avoids, at the expense of its generality and its ideological overtones, the drawbacks of some of its alternatives. It avoids the emphasis on purely technical and formal definition, as in "digital" or "electronic" media; the stress on a single, ill-defined and contentious quality, as in "interactive media"; or the limitation to one set of machines and practices, as in "computer-mediated communication" (CMC). So, while a person using "new media" may have one kind of media technology in mind (such as the Internet), others may mean something else (such as digital TV, cellular telephony, and so on) (Lister et al. 2003, 11).

The new media technologies are seen as extending the interactive possibilities of the old media. For instance texting of radio presenters and the online editions of newspapers contribute to a more inclusive and engaged media audience.

The characteristics of new media marked out above are often linked to the potential for actualizing the tenets of democracy in the lives of the citizenry. It is these features that most proponents of the benefits of the so-called cyber-democracy point to. But what are the visions of African democracy that would be best served by new media technology?

Theorizing Democracy in Africa in the New-Media-Technological Age

One has to be cautious in one's allusions to the concept of democracy, largely because it is an imprecise concept and tends to be used relatively.

To understand this, all we need to do is refer to the definitional debates on "democracy." Scammell and Semetko (2000, xx–xlix) list, and attempt to define, seven variants of democracy as practised across the world, from the West to the socialist republics of the former Soviet Union, including Africa and present-day China. Here, we simply list them to demonstrate the variegated nature of the democratic experience: (1) Direct democracy (socialism, resembling Athenian democracy); (2) Competitive elitist democracy (a critique of socialism and classic liberalism, positing elites as democratic brokers); (3) Pluralism (recognizes competing interests in the body politic), and Neopluralism (acknowledges the power relations inherent in competing interests in the polity); (4) The New Right and libertarianism (emphasizes individual liberties and freedoms and the reduced role of state agency); (5) Participatory democracy (The New Left, emphasizing the processes of participation); (6) Deliberative democracy; and (7) Communitarianism (the elevation of the sense of community and the corresponding *deliberative* edge over other forms of democracy).

The democratic-communicative experience in Africa can be analyzed in terms of three theoretical models, each of which, occurring at specific historical moments, reflects some of the features of the seven varieties of democracy listed above. These theoretical trajectories include the diffusion/modernization, the dependency-dissociation, and the participatory-communication models. It is important to mention here that discussions of democracy and development have generally been intertwined in the African context. Sen (1999, 6), for example, correlates freedom and development, examining "the connection between political and civil rights, on one hand, and the prevention of major economic disasters, on the other." Such a conflation of political and economic rights is clearly evident in much theorizing about participatory approaches to communication (and media) (Servaes 1996). For example, McQuail's notion of the "democratic participatory" role of the media focuses on media's "centrifugal" tendencies toward diversity, plurality, and change (McQuail 1987, 94–96).

The Diffusion/Modernization Model

The diffusion model, fashioned after the "modernization" approach to social progress, was represented in the main by such scholars as Rostow (1960), Rogers (1962), and Lerner (1958), who posited development communication as an engine of change from the "traditional" to the "modern" society. In this regard, "a nation became truly modern and developed when…it closely resembled Western industrial nations in terms of political and economic behavior and institutional attitudes towards technology and innovation, and social and psychic mobility" (Fjes in Melkote 1991, 38).

The model stressed three elements. First, the "psycho-sociological" element entailed "empathy," or the capacity to see oneself in the other fellow's situation. There was a correlation between the expansion of economic activity being equated with "development" and a set of "modernizing" variables, chief among which are urbanization, literacy, mass media use, and democratic participation (Lerner 1958). The second element highlighted the mass media as an "institutional" nexus of modernizing practices and institutions in society, functioning as "watchdogs," "policymakers," and "teachers for change and modernization" (Schramm 1964; cf. Rostow 1960). The third posited "technological" advances as pivotal to the growth of productive agricultural and industrial sectors, and therefore the transfer of technical know-how from the developed North was seen as extremely crucial for development in the Third World nations (Melkote 1991, 24–29).

It is this diffusionist view that sees the media as agents of modernity. Under this logic, the media can serve to "diffuse" those modernizing values required for traditional societies to graduate into modern democratic societies (Rogers 1962). It is here that the media is traditionally seen as a vehicle for modernist values of liberal democracy, positing it as the "watch dog" over state excesses. It is assumed that such democratic conditions are likely to facilitate the process of development (cf. Norris 2006).

The Dependency-Dissociation Model

The dependency model mounted a severe critique of the modernization paradigm, mainly influenced by South and Central American social movements. The "dependistas," as they became known, examined relations between the First and Third Worlds in terms of relations of exploitation and dependency. According to them, "development," in fact, leads to the underdevelopment of the Third World periphery and the concentration of wealth in Western metropoles. Dependency theorists, however, were unable to offer any solutions to the problem of exploitative international economic relations and their impact at micro levels (Tomaselli [n.d.]), but they certainly provided ample material for the participatory-communication model.

This view would treat new media technology as a continuing phase of Third World dependence on Western forms of technology and ways of organizing social and political life. It would thus call for a de-linkage from such dependence and urge the exploiting of indigenous forms of technology and mediation.

The Participatory-Democratic-Communication Model

This model is more than inclusive as it emphasizes the importance of the cultural identity of local communities, as well as democratization and participation at all levels. The concepts of participation, cultural identity and empowerment, and also of the Freirian notions of dialogical communication are central in both the process and context (Servaes [n.d.]).

"Another" communication thus favors what McQuail (1987, 97) refers to as "multiplicity, smallness of scale, locality, deinstitutionalization, interchange of sender-receiver roles and horizontality of communication links at all levels of society." In a sense, this is a shift from the positivist-instrumentalist approach of the modernization paradigm toward a less quantitative and more qualitative and normative model (Melkote 1991, 234).

As an extension of Freire's dialogical pedagogy, participatory communication becomes a process of "conscientization" in which dialogue is both more receiver-centered and more conscious of social structure. Freire (1996) argues that in the traditional pedagogical systems, the receivers were supposed to be uncritical and passive, ingesting the worldview of the elites and then perceiving their problems and needs in terms of the elite-dominated rationality. He called for a new dialogical pedagogy in which the receiver would be liberated from his/her mental inertia, penetrate the ideological mist imposed by the elites, and perceive the realities of his/her existence. It is within the context of this conscientization that theory can be appropriated as praxis for social and political transformation. Participatory communication is thus predicated on the notion of "empowerment."

Rensburg (1994) sees "empowerment" as a move to inform and motivate the community to advance development. Participatory communication entails a great deal of emphasis on what he calls "grassroots participation." In that sense, argues Rensburg, participatory communication tends to be pluralistic and does not suffer from the authoritarian overtones of the dominant paradigm. It enables the community to set its own priorities and standards, which may be unique to its problem situations.

How do these theoretical visions of media and democracy in Africa equate with the new media technology debate? First, they mirror the *theoretical* approaches to the democratic potential of new media. For example, the two "dreams" of "cyber-democracy" that Hagen (1997) identifies are reminiscent of the tripartite theoretical schematization presented above. The first is the more conservative and libertarian vision of cyber-democracy, stressing the importance of a free market and unfettered capitalism. This is the modernization approach to new media, seeing them as vehicles for economic growth and its democratic spin-offs. The second is the more liberal and communitarian vision, which privileges community values (cf. Poster 2000). This is much closer to the participatory-communicative vision. The idea of participatory (democratic) communication is, in a sense, a *dissociation* from *dependency* upon the merely diffusionist vision of media. Seen as such, it can be argued that African notions of new media technology assume a degree of dependency-dissociation, privileging more culturally and socially rooted appropriations of new media technology. This is evidently a less technologically determinist approach and assumes a reasonable degree of social shaping in the appropriation of new technology.

Second, and more importantly, the participatory-communication model seems to be consistent with much literature on the *practical* uses of new media technology in the democratic project. For example, Ogden views cyber-democracy as "the exercise of democratic principles in cyberspace" (Ogden 1996, 128). It seeks to exploit the "new" benefits of computer technology, as highlighted earlier. In particular, it seeks to exploit the interactive nature of new media to enhance citizen participation. As Ogden suggests, it implies "an electronic form of grassroots direct democracy beyond that of local ballot initiatives and referenda" (Ogden 1996, 128).

Bucy and Gregson (2001, 358), in reflecting on "new media use as political participation," argue that this emergent form of electronic democracy or e-democracy (a type of political participation *through* media) involves not just net activism but also the broader range of citizen actions that can take place online, over the airwaves, and through exposure to political messages—actions that invite involvement. These actions include, but are not limited to, direct leader/legislator contact, public opinion formation, participating in civic discussions and agenda building, mediated interactions with candidates and other political actors, donating to political causes, and joining mobilizing efforts—each of which may contribute to the psychological feeling of being engaged with the political system. In a nutshell, then, new media technology promises to transcend the limitations of modern forms of political organization, with a view to broadening such forms for enhanced citizen involvement and participation. This argument for the social shaping of technology (Williams and Edge 2001) leads Barber to rightly ask the question: "which technology and which democracy?" (Barber 2003, 33).

There seems to be general agreement that the kind of democracy being advocated is certainly an improvement on the liberal one. At best, according to Barber, liberal representative democracy is "thin" democracy, in which representative institutions dominate and citizens are reduced to mere "monitors" of the system (Barber 2003, 36). He prefers "strong democracy," one that, though not necessarily always direct, incorporates strong participatory and deliberative elements. As noted earlier, this model postulates the engagement of citizens at the local and national levels in a variety of political activities and regards discourse, debate, and deliberation as essential conditions for reaching common ground and arbitrating differences among people in any given large, multicultural society. In strong democracy, citizens actually participate in governing themselves, if not in all matters all of the time, at least in some matters some of the time (Barber 2003, 37).

The question by Barber—which technology and which democracy?—assumes that the type of democracy we desire will dictate the type of technology to be employed. This is a good assumption because it avoids the McLuhan (1964) technological determinism. The question invites us to think critically about the democracy we seek and the technology we need to use to attain our goal. This leaves room for the deployment of a multimedia (a mix of both old and new media) strategy in our democratic project.

Morrisett (2003, 26–30) offers a vision of the types of technological features that could enhance democratic freedom. To him, there are six important elements required to define democratic uses of new interactive information technologies, including the Internet. These are access, information and education, discussion, deliberation, choices, and action. He is clearly drawing upon the deliberative-communitarian-participatory model of democratic politics, extendable to the participatory-democratic-communication model discussed above. As such, they are applicable to Africa, at least in theory. The next section serves to contextualize the applicability of these new media elements in Africa.

A Critical Review of New Media Use in Africa

Access

Citizens hunger for access to their leaders and to means for expressing their own opinions and judgments. It is this problem of access for a growing, diverse, and dispersed population that makes a national system of interactive information technology (IT) civically useful. If well designed, such a system could counter divisive trends and help bring the nation together.

While universal access is usually an admirable political goal of almost all governments in Africa, it is not easily realizable. Nulens (1997) identifies three key subproblems associated with the problem of access: (i) operational; (ii) contextual; and (iii) strategic. *Operational* problems relate to the day-to-day technical problems that bedevil the implementation of new technological solutions, such as power or electrical deficiencies and inaccessibility of transmission channels, such as satellites.

Contextual problems refer to the cultural environment within which new technology plays out, such as apprehension that the transfer of Western technology only leads to economic and cultural dependency. In other words, technology is not neutral, and ICT policies must thus take into account the potential sociocultural problems. Such cultural issues may buffer the appropriation of new media technology.

Strategy problems reflect the political economy of communication. They arise out of power struggles over economic and other forms of control. Examples of these would be some of the telecommunications transnational companies whose business (profit-making) interests may go against the national-developmental aspirations in Africa. Such companies may influence international policymaking institutions on ICT matters (Nulens 1997, 6). Another example could be the protracted wrangling among African countries, especially Kenya and South Africa, over the New Partnership for Africa's Development (NEPAD)'s proposed Eastern Africa Submarine System (EASSy) project to roll out fibre-optic cables from South Africa to Sudan (allAfrica.com 2006). The problem concerned the type of financing model—private or state or mixed—to be chosen. It was also related to who should have ultimate control over the project. In addition, there was concern as to which private company should be contracted to undertake the construction of the cable.

Information and Education

A vital part of any deliberative discussion is the provision of relevant information and education or knowledge. In the absence of such information and education, debate is likely to be based on opinion rather than fact, prejudice rather than knowledge (Morrisett 2003, 28). An interactive IT system itself need not contain the vital information, as references could be made to other sources of information, from material in libraries to TV documentaries.

Much newly mediated content in Africa leaves a lot to be desired. At the same time, there is also cause for celebration. While the quantity of online content has increased over the years, including that in vernacular languages such as Kiswahili, the educational quality of some of the content is doubtful. In many cases, such information is not *knowledge*, or information mixed with contextual experience. All the same, a good number of Africans find many of the online resources extremely useful. For example, in terms of cellular telephony, it is clear that its educational and informational quality is increasing by the day. Content providers, such as broadcasters and banks, are entering into strategic partnerships with distributors, such as cellular telephony companies, in a bid to introduce client-oriented services. While this may be said to be a financial information service, it still represents a democratic expansion of the communicative sphere in the sense that it has potential to result in a greater number of media platforms. There is simply no telling what other social uses new media might be put to (cf. Banda 2006).

At the same time, there is anecdotal evidence to suggest that new media technology has had a dystopian usage. Indeed while the use of cell phone text messaging as a platform to distribute ethnic hatred in Kenya during and after the 2007 general election (Abdi and Deane 2008) may not be taken as a general trend, it portends something more sinister. These developments clearly demonstrate the regulatory problems likely to emerge as African governments continue to encounter the challenge of new media technology.

Discussion

IT could stimulate discussion not only between citizens and their leaders but among the citizens themselves. Unlike traditional broadcasting, networked computing, whether through conference or e-mail, encourages people to communicate with one another (Morrisett 2003), more broadly and intensively than without the system. Morrisett concludes that electronic technology can be used to provide a modern-day equivalent of the barbershop, connecting citizens across the nation and with their leaders.

In the African context, there is anecdotal evidence of the Internet connecting ordinary people with some centers of power, including mainstream media. For example, Goldfain and Van der Merwe (2006, 120), in their study of the political role of "web-blogging" in Johannesburg, concluded that although blogs are not well established in South Africa, their function is to "provide citizens with an alternative source of news, add more perspectives to the events and issues of the day, and initiate conversation." A "blog"[1] serves as "an aggregator of information… [and] …has the potential to give minorities a voice." This seems to chime with some earlier research findings by Spitulnik (2002), who contends that the existence of the Internet has enabled a dramatic proliferation of completely new ways of communicating, networking, forming community, and maintaining diasporic identities. She gives examples of such electronic forms as electronic mail discussion lists (listservs), Usenet groups and web-page guest books, and chat forums (Spitulnik 2002, 186).

Spitulnik cites the case of the Zambian listserv, Z-list. She reports that the list is used primarily by Zambians living abroad, mostly men, for political discussions and news exchange. The list carries an average of 40–60 postings per day. Debates, political commentary, humorous jabs, and friendly in-jokes are intensive, and some threads continue for weeks with scores of postings. Occasionally non-Zambians post on the list, but for the most part, the list is dominated by a very tightly knit community of Zambian nationals who have built up a history of friendly rapport and repartee (Spitulnik 2002, 187).

Other listservs have emerged from within Zambia, such as the Zambia Media Forum (The-Zambia-Media-Forum@googlegroups.com), connecting members of the journalistic and allied professions around some core issues. Even at such a miniaturized level, one discerns the near absence of women, raising questions about the extent to which new technology is gendered, compelling Plou (2003, 16) to lament that "the 'new' ICTs already reflect many of the gender patterns (in relation to power, values, exclusion and so on) that have been evident for decades in relation to the 'old' media."

Online discussions reflect the institutional settings in which subscribers find themselves. For example, as a member of the Zambia Media Forum, one of us found out how institutionally constrained some members of the Forum were. They could not, for example, append their signatures to a press statement prepared to support the decision by the Government of the Republic of Zambia to raise the mining tax and royalty regime. They feared that they might jeopardize their institutional standing. Some institutions have clear policies about the kinds of social causes their employees or members can support. As such, it is evident that institutional affiliations constrain the extent and nature of participation, contradicting the supposed open democracy embraced online. An important conclusion, then, is that online discussions reflect the discursive practices in which discussants are located and, as such, carries along with them all the structural rigidities, stereotypes, and practices of everyday terrestrial life.

Deliberation

Any interactive communication system must provide the means for deliberation, that is, the careful consideration of an issue and the likely consequences of decisions. For deliberation to occur, provision must be made for the presentation of various sides of a question and attention given to different approaches to outcomes. Examples abound of the "interactivity" of the communication processes characterizing most of the Internet-based communicative forms (cf. Tambini 1999, 315). In elevating this quality, there is an implied criticism of old media as providing a largely unidirectional system of communication, based on formulaic entertainment-based media genres, with audiences treated as mere receivers of mediated content.

Barber (2003, 38) has a word of caution, however. Deliberation can be held back by the type of technology deployed. In elucidating his misgivings, he points to the elements of "speed, reductive simplicity and solitude" characterizing new technology as potential dangers to a deliberative form of democracy. Barber sees "speed" as both the greatest virtue and the greatest vice of digitalized media. Digital media are in a rush. According to Barber, the impact of "fast" varies, however, depending on the version of democracy we postulate. With representative democracy, for example, accelerated pace may make little difference, at least for citizens. Where thought and deliberation are not essential, a speeded up political process may simply appear as time-saving and efficient. For deliberation to occur, Barber warns us, there is need for new technology to slow down, although he admits that it would be like asking a hare to run at the same pace as a tortoise!

Barber also laments the possibility that digital media, already implicated in the binary dualisms of "on/off" and "zero/one," might reduce citizens' participation in politics to the barest necessities of voting between polarized alternatives, without subjecting the available choices to serious political deliberation (Barber 2003, 39).

In the same vein, Barber (2003, 39) argues that digital media encourage "a politics of solitude" in which citizens just sit "at home in front of electronic screens and view the world and its political choices as so many consumer alternatives." He dismisses the countercriticism that refers to "virtual communities" as merely pointing to "vicarious conglomerates lacking the empathy and need for common ground that define real-world communities" (Barber 2003, 39).

In Africa, the qualities of "speed, reductive simplicity and solitude," while perhaps true at an interpersonal level, are clearly not true at the societal level. This, in part, is due to the low levels of penetration of new technology in the majority of African countries. Outside of institutional settings, not many people can afford the luxury of sitting in front of a computer all the time, typing away on the keyboard. Nor are TV sets equipped with interactive capabilities to allow for speedy interchange. Another problem is located in the political structures of most African countries. Political decision-making is centralized in a coterie of ruling-party stalwarts who are often averse to the decentralizing tendency of new media technology. As a result, the supposed revolutionary effect of new media is caught up in the rigid structures of organizational and political decision-making as well as in the daily economic struggles of the people.

Choices and Action

Discussion and deliberation are sharpened when participants understand that choices among alternative courses of action must be made. An interactive system devoted to e-democracy needs to be organized in such a way as to clinch choices about some course of action. In other words, users of new technologies must understand that when they go through the hard work of education, discussion, and deliberation, their choices and judgments will be used (Morrisett 2003, 29–30), or acted upon.

It is not entirely clear whether every involvement in cyberspace must result in specific actions. However, a body of empirical evidence is emerging across Africa suggesting that such involvement does result in some or other form of action. An example of collective action engendered by the Z-list is documented by Spitulnik. Thirty-five Z-listers signed a letter to the editor of the *Post* on the severe deterioration of academic standards and academic freedom at the University of Zambia (UNZA). Using the UNZA situation as a microcosm of a crisis in academe across the nation, the authors of the letter elaborated several areas of concern (for example, resource development, corruption, low morale, and staff retention) and suggested "policy initiatives that can correct that situation" (Spitulnik 2002, 188).

Although there was no specific governmental reaction, it is clearly an example of how online communication could result in certain forms of democratic organizing and mobilizing. It is an example of how alternative public spheres are structured around communities of ideological interest, even though sustained action may not be the intended outcome. In a sense, this evidence can be used as a critique of the deliberative, action-oriented approaches to democracy. The media fact of *communicating* is an action. In fact, one could suggest that it represents a form of *deliberation*, as people learn to talk to one another (cf. Mouffe 2000, for an excellent treatise on "agonistic pluralism" as an alternative to deliberative democracy).

Interrogating New Media and the Public Sphere in Africa

The contextualization of new media technology above provides the basis for the empirical analyses in the ensuing chapters, which explore specific dimensions of the impact of new media on African social formations, interrogating the extent to which the celebratory, cautionary, and dystopian perspectives are borne out by the evidence relating to ICTs and the enhancement of democratic practice. The overarching question that they all examine is whether new media technologies do *indeed* improve and qualitatively change the continent's political structures, systems, and processes, or only create a quantitative increase in access to information and interactions stemming from the ease and relative cheapness of communicating via new media technology (Polat 2005). These chapters are, thus, concrete tests of Polat's observation that the Internet, as a communication medium, increases the communication capacity in an unequal way by supporting some forms of communication more than others and, by extension, creating asymmetrical forms of political participation. The authors provide sophisticated and nuanced analyses of these issues in the African context as various actors appropriate new media technologies in defined spheres of life.

The next two chapters focus on ICT use by governments and those seeking political power. Using the related concepts of e-government and e-governance (e-gov), Mudhai examines, in chapter two, the democratic implications of government use of ICTs, and its associated challenges, especially in South Africa and Nigeria. He interrogates the extent to which the employment of e-gov in these countries is enhancing efficiency, effectiveness, transparency, as well as citizen inclusion and participation in democratic politics. The chapter not only engages with the challenges of e-governance initiatives but also highlights their contribution to good governance. This discussion is followed, in chapter three, by Nyabuga and Mudhai's analysis of efforts by Kenyan political parties, and their presidential candidates, to join the new Information Age, especially around elections. They argue that while new media may have some potential to help monitor and mobilize political activity and possibly encourage political engagement, they can also reinforce the positions of those in power.

The next seven chapters examine how elitist as well as marginalized players in domestic civil society use ICTs for sociopolitical, as well as economic-developmental, purposes. These actors, who operate transnationally, have become versatile political agents over whom states have increasingly less control. Chapters four to six examine the activities of mainly elitist, urban civil society organizations and NGOs in the domestic public sphere. In chapter four, Moyo discusses how news media institutions, civic organizations, and ordinary citizens in Zimbabwe are using the Internet and cell phones in information-gathering, dissemination, and presentation to promote democracy and human rights. This chapter also gives close attention to new media infrastructure, the regulatory environment, and quality of service provision for Internet and cell phones, and to other challenges facing new media users in Zimbabwe. Frère and Kiyindou focus on francophone Africa, in chapter five, arguing that even though the ICTs have spread in an environment already transformed by political, media, and civil society liberalization, they have nevertheless brought additional changes. This chapter shows how the Internet has enhanced citizen engagement in the public sphere through increasingly critical public debates, more access to official information, and more interaction of local civil society organizations with their counterparts abroad, thereby giving African citizens more capacity to participate in local and worldwide public spheres. The authors contend that these developments notwithstanding, there is concern that the technology may be facilitating the erosion of African values and their role in building democratic institutions that are appropriate for the African context. In chapter six, Ayyad discusses how NGOs, in Egypt, have used their Web sites to examine serious issues and obstacles confronting socioeconomic and political development in that country. These issues include corruption, human rights, and democratic expression.

Wakunuma-Zojer and Litho examine, in chapter seven, the extent to which the rhetoric of "empowerment," in particular for women in traditional societies, is being transformed into practice on the ground. The authors take a close look at the political, social, and economic impact of the Internet and cellular telephony on the "empowerment" of rural women in Uganda and Zambia. They argue that ICT projects targeted at women in these countries, and other developing societies, often privilege the Internet at the expense of other ICTs, yet cell phones are being appropriated and used at an exponential rate in these communities. Dralega contends, in chapter eight, that there is an urgent need for the provision of communicative platforms with which rural youth can face mainstream discourses that overlook minority concerns. This assessment is based on a critical review of ICT initiatives that aim to revitalize rural youth's engagement in Uganda's public discourses.

Another dimension of the ICT-public sphere matrix takes into account the opportunities that have emerged for transnational networking of deterritorialized citizens to engage with the politics of "home." In chapter nine, Tettey explores the extent to which Africans in diaspora settings utilize their agency as transnational citizens, within the mediascapes made possible by the Internet, to engage in political discourses about their home countries, pursue long-distance nationalism, and attempt to shape politics and public policy in their countries of origin. This is followed in chapter ten by Frenzel and Sullivan's critical assessment of how a digital activist network, Indymedia Africa, spreads its organizational principles into the African context, creating spaces for virtual and physical convergence by transnational actors. The case highlights the technological and pragmatic limits of (new) media as avenues for transnational collaboration and solidarity, as actors and ideologies clash in these socially constructed spaces.

The next two chapters explore ways in which popular culture makes use of new media to distribute messages of political significance to the general population. In chapter eleven, Tomaselli and Teer-Tomaselli provide an analysis of how ordinary people, politicians, civil society, and religious organizations have appropriated and responded to "Reality TV" programs for political ends, with a special focus on *Big Brother*. This is followed in the next chapter by Ogola, Schumann, and Olatunji's argument that although African governments try to monopolize or manipulate public spaces for popular expression, new spaces of freedom and dissent have been forged and fashioned around popular cultural forms. They demonstrate how musicians take advantage of new media tools to influence national politics in Nigeria, Côte d'Ivoire, and Kenya.

The last two chapters focus on the relevance of ICTs in conflict or post-conflict environments. Frère argues, in chapter thirteen, that in the midst of hostile conflict environments in the Democratic Republic of Congo, Rwanda, and Burundi, journalists have seen their professional practice greatly transformed by the new technologies of information. She observes that ICTs have helped develop a better understanding, by local journalists, of the complex conflict situation in their countries, thereby enabling them to make significant contributions to peace-building efforts in the Great Lakes region. The volume concludes with Ottosen and Mudhai's discussion, in chapter fourteen, of the possible consequences of the use of new media (on public opinion, attitudes, and behavior) in environments of ethno-political tension and conflict, especially violent ones. They also explore the opportunities provided by the new media platforms for propaganda and counterpropaganda, as exemplified recently in Kenya, Zimbabwe, and South Africa.

From the foregoing analyses, there is no doubt that new media technologies have a place within the African democratic scheme, but their oft-cited prospects must be interrogated in terms of the economic, political, and cultural realities of the continent. Following Barber (2003), it is clear that Africans need to strike a cautionary note about the supposed emancipatory nature of new media—technology must be tamed to serve the specific type of deliberative, participatory democracy Africa may need at this juncture in its political development. This cautionary note is even more important now, given the Kenyan hate messages by cell phone (Abdi and Deane 2008, 10). More importantly, however, this reminds us that it is human minds that must think about the shape of technology. This view accords with the social shaping of technology perspective as a way of negotiating with McLuhan's technological determinism.

Notes

1. A blog is a periodic, and often continuously updated, Web site that posts the thoughts and observations of a single writer, and often the responses to those observations.

References

Abdi, J. and J. Deane. 2008. *Policy Briefing # 1: The Kenyan 2007 Elections and their Aftermath: The Role of Media and Communication*. London: BBC World Service Trust.

Adeya, C. N. and D. L. Cogburn. 2001. "Globalization and the Information Economy: Challenges and Opportunities for Africa." In *The Digital Divide in Developing Countries: Towards an Information Society in Africa*, ed. G. Nulens, N. Hafkin, L. Van Audenhove, and B. Cammaets, 77–109. Brussels: Vub University Press.

AllAfrica.com. 2006. *East Africa: EASSy Project to cost $280 Million*, http://allafrica.com/stories/200608290061.html (accessed September 8, 2006).

Appadurai, A. 1996. *Modernity at Large: Cultural Dimensions of Globalization*. Minneapolis: University of Minnesota Press.

Banda, F. 2003. *Community Radio Broadcasting in Zambia: A Policy Perspective*. PhD thesis, University of South Africa, Pretoria.

———. 2006. *Zambia African Media Development Initiative (AMDI) Research* Report. London: The BBC World Service Trust.

Barber, B. R. 2003. "Which Technology and Which Democracy?" In *Democracy and New Media*, ed. H. Jenkins and D. Thorburn, 33–47. Cambridge, MA and London: The MIT Press.

Bucy, E. P. and K. S. Gregson. 2001. "Media Participation: A Legitimizing Mechanism of Mass Democracy." *New Media and Society* 3(3): 357–380.

Fair, J. E. 1989. "29 Years of Theory and Research on Media and Development: The Dominant Paradigm Impact." *Gazette* 44: 129–150.

Fourie, P. J. 2001. "Globalization, the Information Superhighway, and Development." In *Media Studies. Volume 1: Institutions, Theories and Issues*, ed. P. J. Fourie, 593–625. Lansdowne: Juta.

Freire, P. 1996. "Pedagogy of the Oppressed." In *The Political Economy of Development and Underdevelopment*, 6th edn, ed. K. P. Jameson and C. K. Wilber, 551–567. New York: McGraw-Hill.

Goldfain, K. and van der Merwe, N. 2006. "The Role of a Political Blog: The Case of *www.commentary.co.za*." *Communicare* 25(1): 103–121.

Gumucio-Dagron, A. 2001. *Making Waves: Stories of Participatory Communication for Social Change*. New York: Rockefeller Foundation.

Hagen, M. 1997. *A Typology of Electronic Democracy*. http://www.uni-giessen.de/fb03/vinci/labore/netz/hag_en.htm. (accessed September 5, 2006).

Huber, B. 1998. *Communication Aspects of Participatory Video Projects: An Exploratory Study*. http://ictupdate.cta.int/index.php/filemanager/download/236/ (accessed February 5, 2007).

Jensen, M. 2002. African Internet Status. *The African Internet—A Status Report*. http://www3.sn.apc.org/africa/afstat.htm (accessed September 8, 2006).

Landow, G. 2003. "The Paradigm is More Important than the Purchase." In *Digital Media Revisited: Theoretical and Conceptual Innovations in Digital Domains*, ed. G. Liestøl, A. Morrison, and T. Rasmussen, 35–64. Cambridge, MA and London: The MIT Press.

Lerner, D. 1958. *The Passing of Traditional Society: Modernizing the Middle East*. New York: Free Press.

Lister, M., J. Dovey, S. Giddings, I. Grant, and K. Kelly. 2003. *New Media: A Critical Introduction*. London: Routledge.

McLuhan, M. 1996. *Understanding Media: The Extensions of Man*. New York: McGraw-Hill Book Company.

McQuail, D. 1987. *Mass Communication Theory: An Introduction*, 2nd edn. London: Sage.

Melkote, S. R. 1991. *Communication Development in the Third World: Theory and* Practice. New Delhi: Sage.

Morrisett, L. 2003. "Technologies of Freedom?" In *Democracy and New Media*, ed. H. Jenkins and D. Thorburn, 21–32. Cambridge, MA and London: The MIT Press.

Mouffe, C. 2000. "Deliberative Democracy or Agonistic Pluralism." In *Political Science Series 72*, ed. C. Neuhold. Vienna: Institute for Advanced Studies.

Norris, P. 2006. *The Role of the Free Press in Promoting Democratization, Good Governance, and Human Development.* Paper Presented at the Midwest Political Science Association Annual Meeting, April 20–22, 2006. Chicago: Palmer House.

Nulens, G. 1997. "Socio-Cultural Aspects of Information Technology in Africa: The Policy of the World Bank Studies on Media Information and Telecommunications (SMIT)." *Communicatio* 23(2): 15–23.

Ogden, M. R. 1996. "Electronic Power to the People: Who is Technology's Keeper on the Cyberspace Frontier?" *Technological Forecasting and Social Change* 52: 119–133.

Pavlik, J. V. 1996. *New Media and the Information Superhighway.* Boston: Allyn and Bacon.

Plou, D. S. 2003. "What About Gender Issues in the Information Society?" In *Communicating in the Information Society*, ed. B. Girard and S. O. Siochru, 11–32. Geneva: United Nations Institute for Social Research.

Polat, R. K. 2005. "The Internet and Political Participation: Exploring the Explanatory Link." *European Journal of Communication* 20(4): 435–459.

Poster, M. 2000. *CyberDemocracy: Internet and the Public Sphere.* http://a-r-c.gold.ac.uk/a-r-c_Two/print_mark.html (accessed September 5, 2006).

Rensburg, R. 1994. "Community Development: Essential Contribution or Paternalistic Communication?" *Dialogus Online* 1(1). http://www.unisa.ac.za/dept/kom/d11radio.html (accessed December 12, 1998).

Robertson, R. 1992. *Globalization: Social Theory and Global Culture.* London: Sage.

Rogers, E. M. 1962. *Diffusions of Innovations.* New York: Free Press.

Rostow, W. W. 1960. *The Stages of Economic Growth: A Non-Communist Manifesto* Cambridge: Cambridge University Press.

Roszak, T. 2004. "The Cult of Information." In *The Information Society Reader*, ed. F. Webster, 55–61. London and New York: Routledge.

Scammell, M. and H. Semetko, eds. 2000. *The Media, Journalism and Democracy.* Aldershot, Burlington, Singapore, and Sydney: Ashgate, Dartmouth.

Schramm, W. 1964. *Mass Media and National Development: The Role of Information in the Developing Countries.* Stanford: Stanford University Press.

Sen, A. 1999. "Democracy as a Universal Value." *Journal of Democracy* 10(3): 3–17.

Servaes, J. 1996. "Introduction: Participatory Communication and Research in Development Settings." In *Participatory Communication for Social Change*, ed. J. Servaes, T. L. Jacobson, and S. A. White, 13–25. New Delhi: Sage.

_______. n.d. *Development Communication – for Whom and for What?* http://www.unisa.ac.za/dept/press/comca/211/servaes.html (accessed 12 December, 1998).

Spitulnik, D. 2005. "Alternative Small Media and Communicative Spaces." In *Media and Democracy in Africa*, ed. G. Hyden, M. Leslie, and F. F. Ogundimu, 177–205. New Brunswick, and London: Transaction Publishers.

Tambini, D. 1999. "New Media and Democracy: The Civic Networking Movement." *New Media & Society* 1(3): 305–329.

Tomaselli, K. n.d. *Politics of Development*. http://www.sacod.org.za/report/politicsofdevelopment.htm (accessed February 5, 2007).

Wilcox, D. L. 1975. *Mass Media in Black Africa: Philosophy and Control*. New York: Praeger.

Williams, R. and D. Edge. 1996. "The Social Shaping of Technology." *Research Policy* 25: 856–899.

CHAPTER 2

Implications for Africa of E-Gov Challenges for Giants South Africa and Nigeria

Okoth Fred Mudhai

If the key goals of e-government and e-governance (e-gov is hereafter used to refer to both terms) are to enhance efficiency, effectiveness, and transparency, as well as aid citizen inclusion and participation in a democracy, then the giants South Africa (SA) and Nigeria are best-placed to lead the way not only in sub-Saharan Africa (SSA) but also the entire continent. This is because of their populous nature, vast economies, expansive geographical dimensions and the complexity of their political and administrative systems. While a recent UN report based on a global survey singled out SA as an e-gov exemplar in SSA, Nigeria only managed a second-best place in the worst ranked West Africa region. Hinged on the notion that the provision of government services and information through electronic or online means is a crucial way of popularizing information and communication technologies (ICTs) among populations, this chapter examines not only the challenges of e-governance initiatives but also probes the extent to which they could aid or hinder efficacious democracy.

The main purpose of this chapter is, therefore, not to provide a comprehensive coverage of the wide area of e-gov in Africa but to focus on those aspects with implications for politics in general and democracy in particular. The chapter first gives an overview of e-gov and then examines its links with democracy before concluding with policy suggestions.

E-Gov in Perspective

Definitions

The assertion by Bellamy and Taylor (2003, 3) that "little has been written on the information age as it relates to governing" no longer holds true. Academics, practitioners, and students have in recent times addressed various issues around adoption, diffusion, and impact of ICTs in local and central government not only for basic public information and communication but also for more ambitious policy goals that may deliberately or inadvertently increase or hinder public participation in sociocultural and economic realms as well as political discourses.[1] However, much of the literatures acknowledge a lack of consensus on the e-government concept (Prins 2001, 1–2; Chissick and Harrington 2004, 3; Misuraca 2007, pref., ch.1 sec.1.3; Saidi and Yared, n.d., II.2). The interpretations are "varied, and sometimes conflicting" (Misuraca 2007, ch.1 doc.5 of 16). For this reason, it is imperative to briefly define the contradistinction between e-government (e-govt) and the related term of e-governance.

Scholars, the Commonwealth, UNDP, OECD, the EU Commission and the World Bank,[2] among others, offer various definitions (Chissick and Harrington 2004; Misuraca 2007, ch.1 sec.1.3). To Drucker (cited in White 2003), e-governance in particular "is about *choice*...," for both the government and citizens. To David L. McClure (cited in Oyawoye, n.d., n.p.), it is more than just G2C (government to citizen); it also involves business (G2B, B2G, government to business and business to government), other government agencies (G2G, government to government), employees (G2E, government to employee)—not to mention Oyawoye's addition of C2G and E2G.[3] It is therefore "multi-dimensional, multi-actor, multi-level and inter-sectoral..., influenced not only by the ICT-revolution, but mainly by globalization, international economic competition and state transformation" (Misuraca 2007, pref).

In essence, most authors attempt to capture a conceptual or paradigmatic shift from e-govt to e-governance (Misuraca 2007, ch.3 sec.3.5). The "old" model mainly deemed as e-govt is more about e-service delivery (ESD), relatively passive automation, data-processing, internal workings, and back-office functions, while the "new" or "smart" model largely perceived as e-governance is more about relatively active interaction or communication, external workings, and front-office functions (6, 2001, 7; Heeks 2002, 4; Heeks 2004, 1, 2, 12; Chissick and Harrington 2004, 5–6; Misuraca 2007, pref., ch.1 sec.1.3, ch.3 sec.3.5; Oyawoye, n.d.; Saidi and Yared, n.d., sec II.1). Combined, they involve governance "with and of" ICTs (Misuraca 2007, ch.1 sec.1.3), the first bureaucratic and the second innovative, transformative, and citizen-empowering. The terms are complimentary and not mutually exclusive (Misuraca 2007, ch.1 sec.1.3; Oyawoye, n.d.).

> ...although you can have e-government without it evolving into e-governance (as is basically the case now in Nigeria), it would probably be difficult to jump straight to e-governance...(Oyawoye, n.d.).

This therefore calls for sensitivity to "very specific environment and contextual atmosphere" (Misuraca 2007, pref.) within which these concepts are put into practice, with particular regard to the motivations.

Rationale of E-Gov

It is vital to understand "why and how...new technologies are diffusing into government" (Bellamy and Taylor 2003, vii). There are various dimensions: economic, social, and governance (Prins 2001, 4; Theunissen 2001, 202; Bellamy and Taylor 2003, 37 and 66; Krishna and Madon 2003, 2; Misuraca 2007, ch.3 sec.3.2, 3.5; Oyawoye, n.d.). Of greater interest for us in this chapter is the political dimension, which is often a spin-off rather than direct motivation for e-gov.

Theunissen (2001, 202) identifies one of two main reasons for e-gov as, "to enable expansion of the public sphere in a democracy." Citing various sources, Misuraca (2007, ch.1) identifies some e-gov purposes as the following: improve "democratic processes" (EU Commission), "transparency and accountability" (UNDP), and "citizen empowerment" (the World Bank). Oyawoye (n.d.) lists eight specific objectives, including broad-based public awareness and citizen participation, and achieving "digital democracy" by supporting e-mediated exchanges (facilitating dissemination of political information, opinion polling, campaigning, and contributions, and ultimately voting).

Within the three domains of e-gov Heeks (2002) identifies remain the aspects that are of interest to democracy. Of the four examples provided under the first domain of improving processes (e-administration), one—creating empowerment—specifically relates to democracy, elaborated on as "transferring power, authority and resources for processes from their existing locus to new locations" (Heeks 2002, 5). The second domain, connecting citizens (e-citizens and e-services), involves "citizens either as voters/stakeholders from whom the African public sector should derive its legitimacy, or as customers who consume public services" (Heeks 2002, 6). The third domain is about building external interactions.

Another author, 6 (2001, 13–20), proposes four rival "theories" on e-governance that relate to motivations, one of them cautions theory of "technology as totem, fetish and foil in ritualized social conflict"—privileging sociopolitical rather than ICT-driven shaping.[4]

Diffusion of E-Gov in Africa

Although 6 (2001, 9) traces the rise of e-governance to 1970s while Prins (2003, 2) argues that modern e-governance was launched in the US in 1993, Heeks (2002) indicates the concept is not new in Africa, as it can be traced back to early 1960s. In recent years, there has emerged great interest and literature on e-gov in Africa (Misuraca 2007; Okpaku 2007; Oyawoye, n.d.). As with most other aspects of new media diffusion, the literature acknowledges a "digital divide" affecting Africa—and other developing regions—thus resulting in slow diffusion of e-gov (Theunissen 2001, 201; Krishna and Madon 2003: 5; Narayan 2007, 1). Barriers to e-governance diffusion in Africa relate to human (including leadership and strategic thinking), infrastructural, technological (including data systems), material, institutional, and legal factors (Heeks 2002, 1; Misuraca 2007, ch.3 sec.3.5; Oyawoye, n.d.). These internal and external factors have resulted largely in failure of e-gov initiatives in Africa.

Misuraca (2007) reports mixed results from studies of e-govt decentralization initiatives in Senegal's Acacia Initiative (no significant impact), Ghana's modernized chiftaincy (unreliable link), Uganda's DistrictNet (challenges hindering success), and Cape Town's "Smart City" strategy in SA (unaltered racial divide). Heeks (2002, 10) identifies three types of results: "total" failure (e.g., a land-licensing system in SA, due to entrenched interests of powerful groups); "partial" failure (e.g., touch-screen kiosks in rural Northwest SA, removed a year later due to lack of updated or local content); and success (e.g., tax systems in Egypt and Mauritius, and the SA election project).

Most e-gov initiatives fail due to a lack of proper understanding of the issues involved in implementation (Oyawoye, n.d.). Even though to others "the central issue behind ICTs utilization in Africa is not a matter of technology transfer [as] it has everything to do with people's empowerment and society's ability to use the technology as a facilitator for democracy, a tool for universal access to services, opportunities and resources" (Misuraca 2007, ch.3 sec.3.5), a large proportion of blame for failures has been attributed to external factors. To Okpaku (2007), low success rates "possibly lie, at least in part, elsewhere in global strategic, intellectual, cultural and policy assumptions that are eminently questionable." Echoing this, Heeks (2002, 5) argues that for Africa, e-gov "is essentially an imported concept based on imported designs," partly by Western trained civil servants, so one way to move forward would be to justify and understand ICTs "in the context of a broader vision and necessity for e-government in Africa." There is need to strategically narrow design-reality gaps along the "third way" lines of contingency approach that "sees no single blueprint for success and failure [but] recognizes that there are situation-specific factors"—an approach that differs from the "factoral analysis," which focuses on constraints, and Giddens' structuration theory or Callon's/Latour's actor network theory (Heeks 2002, 12). On his part, Oyawoye (n.d.) suggests nineteen specific solutions, most of them commonsensical and universal—based on best practices.

> Diverse as we are, African nations are more similar than different where ICTs are concerned. I would therefore like to see a set of African e-governance best practices developed and maintained . . . The development of home-grown solutions should be encouraged.

Coleman (in UN-DESA 2008, 121) proposes three principles for devising efficacious e-gov strategies for Africa: African ownership; public-private partnership; and regular evaluation of impact. One useful strategy would be to supplement what I would term Internet-governance (i-governance) with mobile governance (m-governance) "in a spoke and hub model . . . helpful in bridging the digital divide . . . " (Narayan 2007, 1 and 13). These could help address Africa's "lack of 'e-readiness for e-government' " (Heeks 2002, 1).

As things stand now, "African countries rank very low on the global measure of e-government readiness" (Kalu 2007)—defined by Heeks (in Misuraca 2007, ch.1 sec.1.3) as "available technological infrastructures, legal frameworks, institutional and human resources and political will." Africa "lagged far behind" in the 2008 UN e-govt readiness survey (UN-DESA 2008, xiii, 19; UN, cited by Perry 2008). It occupied the same "lowest in the world" position as it did in a similar survey in 2005 (UN-DESA 2005, 29). Although "most e-government projects, both in industrialized and developing countries, fail totally or partially" (Misuraca 2007, ch.3 sec.3.2) due to attendant implementation risks, Kalu (2007) suggests that "African leaders" need to address the continent's "low e-readiness . . . by . . . creating an enabling environment." There is need for caution, however, given the sensitivity of politicians to political impact of ICTs as well as the necessity of cost-benefit analysis (cf. Okpaku 2007).

For this reason, deployment of e-gov in Africa should be gradual and cautious. It is, however, worth noting that e-gov initiation and costs need not necessarily be restricted to the government, given that "social computing" activities of civil society organizations and individuals are "positively contributing to realizing the key goals of better, simpler, joined-up and networked government" in the UK (Osimo and Punie 2008, 28, 31). Indeed, "despite the many indicators showing Africa at a disadvantage, the potential for growth through integrating ICTs in the governance systems is encouraging" (Misuraca 2007, ch.3 sec.3.5), with many countries such as Egypt, Ghana, Kenya, Mauritius, Morocco, Nigeria, South Africa and Tunisia, Ethiopia, Uganda, Gambia, Zimbabwe, and Senegal developing supportive national ICT strategies and programs (Misuraca 2007, ch.4 sec.4.2). Many have developed e-government in various areas at the levels of the local and central governments: e-administration, e-health, e-education, e-agriculture, e-customs, e-voting, e-policing, e-taxation, and so on. In spite of a country like Kenya being still at the stage of automating back-offices (Okong'o 2005), it has a Directorate of E-Government (www.egovernment.go.ke). It is, however, the e-gov experiences and challenges of South Africa and Nigeria that has ramifications for the "movement," especially across SSA due to their standings on the continent.

The populations of SA (around 50 million) and Nigeria (around 150 million) as well as their sizes and influences make them Africa's India and China.[5] SA has relatively more resources than its neighbors—including in the realms of ICT (Parkinson 2005: Ch.2), which makes it "a standout e-government leader" in SSA (UN-DESA 2008, 26; Perry 2008). SA's "E-Government Access Strategy 2003–2008" shows clear focus areas (White 2003). In 2008, SA's e-govt readiness ranking was the highest in Africa whereas Nigeria was placed second to Cape Verde in the lowest-ranked West Africa region (Perry 2008; UN-DESA 2008). Although "Nigeria's immense and diverse population…and its wealth of natural resources make it a microcosm of Africa" (Okafor 2008, blurb), its e-gov practice is "negatively impacted upon" by its "poor state of social infrastructure (especially power supply and road network)" (Dode 2007). All the same, Nigeria has made some strides in the much-needed local contextualization of e-gov.

> In Nigeria, NITDA [National Information Technology Development Agency] has developed a (software) Nigerian keyboard that can generate local symbols and allows for the easier use of local languages on computers. All organs of the federal government have also been directed by President Obasanjo to make the only company (at that time) manufacturing computers locally to an international standard, the supplier of first choice for computer systems… (Oyawoye, n.d.).

Through another private sector intervention coupled with government commitment, the Digital Knowledge Democracy was launched in three Nigerian cities (Abuja, Enugu, and Lagos) in late 2008.[6] Most African countries are still struggling to make the paradigmatic shift from e-govt to e-governance, but this has not stopped them—especially the giants SA and Nigeria—from making an effort in that direction, with implications for politics in these countries and around Africa.

E-Gov and Democracy in Africa

Most political leaders, especially in Africa, embrace e-gov with the intention of enhancing government efficiency, including cost-saving, and pleasing donors rather than empowering citizens. Democratic outcomes are often spin-offs of administrative and economic imperatives. Although the impact on politics may be questionable (Okpaku 2007), being "more a wishful thinking than a proven reality" (Misuraca 2007, ch.1 sec.1.3), a number of scholars indicate that the link between e-gov and democracy cannot be wished away. "Better governance, thanks to the ICTs, would improve, according to Okot-Uma, democracy and ultimately peoples' lives" (Misuraca 2007, ch.1 sec.1.3), given the potential of ICTs "as catalysts and enablers for democratic governance and the promotion of democratic practices . . . " (Theunissen 2001, 202). Bellamy and Taylor (2003, 63) point out that one "fundamental" question with regard to e-gov "is whether the [often primary goal of] enhancements in efficiency and effectiveness . . . are capable of creating a political momentum sufficient to overcome constitutional objections and political resistance. . . . " Stressing the need for certain preconditions, Saidi and Yared (n.d., sec.1.2) argue that apart from economic benefits, e-govt "as a major instrument for achieving 'good governance' . . . *can* be an important source of . . . democracy" as long as it is "efficiently and purposefully implemented" and is "accompanied by important investment. . . . "

One of the key assumptions of e-gov, from Stephen Clift's definition of e-democracy (Misuraca 2007, ch.1 sec.1.3), is about it resulting in "more empowered citizens influencing policy priorities" (Prins 2001, 1), about the "increasingly powerful" nature of the "i" and "c" in ICTs as "the power of ICTs lies in their proleptical vision" (Bellamy and Taylor 2003, 63). This includes the "hugely potent claim that liberating the power of new technology" revolutionizes government-citizen relations—"changing the terms of trade-off between efficiency, effectiveness, quality and democracy" (Bellamy and Taylor 2003, 65).[7] While Bellamy and Taylor (2003, 65–66) concur that "technologies *can*, undoubtedly, be used to alter the trade-offs between competing values in government," their own "analysis . . . show that the promise of ICTs is not as unambiguously benign as the advocates of reinvention suppose," given that "whether, how and where those trade-offs are made is, in practice, influenced by many factors, including political and bureaucratic motivations . . . " The duo identify three archetypes of e-democracy or teledemocracy, "strong" (enabled by low-cost access to ICTs) or "populist" (top-down "teleocratic" control) or "consumer" democracy, which can result from the deployment of ICTs in governance, mainly depending on citizens' level of engagement—"active" or "republican," and "passive" citizenship (Bellamy and Taylor 2003, 91–118). While consumer democracy—manifested through "bureaucratic processes and service delivery" with such aspects as "democratic intelligence" (information gathering by government), "managerial democracy" (competence and information base of public servants and party managers), and "telefascism" or "hi-tech totalitarianism" (by isolated individuals)—could "simply augment and speed up the decentring of representative democracy,...the potential exists for the use of ICTs to 'recentre'politics" (Bellamy and Taylor 2003, 117).

> For while we argue that the "reinforcement of politics" thesis provides a powerful explanation of current trends, it begs the question of what character of politics is being reinforced (Bellamy and Taylor 2003, 117).

This link between *realpolitik* and e-gov is as true in Africa as it is elsewhere, hence various social, economic, and political factors determine whether e-gov is accepted and implemented, as well as how far it affects democracy. Recent events in Kenya and Zimbabwe show that democratic outcomes cannot be attributed solely to information exchange, considering the role of factors such as violence and intimidation. All the same, given the largely oral nature of African populations, of crucial importance is the potential of cell phones, "in conjunction with other converging technologies, to be one of the strongest catalysts in maintaining and facilitating democracy and democratic processes" (Theunissen 2001, 210).

Linked to the e-democracy aspect of e-governance is the concept of e-participation in so far as "some measure of power" goes to the citizen from "the political and bureaucratic elite" (Misuraca 2007, ch.1 sec.1.3). The UN definition of e-participation emphasizes the process of information-consultation-decision making (Misuraca 2007, ch.1 sec.1.3; UN-DESA 2008, 17). Africa fairs badly in the UN e-participation index (UN-DESA 2008, 58, 60).

The fact that most African countries are ranked at the bottom globally (UN-DESA 2008, Table 8, p. 212), with only four at the top fifty, shows that most governments in the continent have yet to enhance their "ability to request, receive, and incorporate feedback from constituents" in order to tailor "policy measures...to meet the needs and priorities of citizens" (UN-DESA 2008, 58ff). Under e-information (the extent to which Web sites and portals provide basic information), Congo, Egypt, and Togo use e-mail to update their citizens, and Togo uses RSS feeds to update and involve citizens. Under e-consultation, Botswana is one of the developing countries that scored globally a position among the top twenty-five nations that employ interactive methods to solicit citizen opinion, feedback, and input, such as online channels, including informal polls, bulletin boards, chat rooms/instant messaging, and web-logs (blogs), as well as formal online consultation; Botswana is listed with Cameroon, Congo, Ghana, Mauritania, and Mozambique among countries that use open web forum for discussing topics. Under e-decision-making, few African countries definitely acknowledge individual citizen input and commit to take it into account when making decisions;

Mozambique is listed as the only African country that publishes findings/results of citizen opinions, including e-opinions, on Web sites.

It is worth noting that e-govt readiness ranking, and income level, does not necessarily correspond with e-participation as one may be tempted to expect. Countries in the top eleven e-govt readiness rankings in Africa, such as Kenya, Algeria, and Tunisia, are in the bottom twenty for e-participation rating, while those with very good e-govt readiness ranking, such as Mauritius and Seychelles, have just average e-participation rating. Mozambique, with a poor e-govt readiness ranking, tops the e-participation rating, while Burkina Faso with very poor e-govt readiness rating is a decent fifth in e-participation. Nigeria's rating shows it is making better use of its limited e-govt status in terms of e-participation. All the same, Angola, Egypt, Guinea-Bissau, Libya, and to some extent Ethiopia and SA have their rankings corresponding in the two indexes. This shows that e-participation and e-democracy aspects of e-gov are more than just issues of access, including in relation to richness—given that a low income country tops Africa's e-participation ranking while a number of lower middle-income and upper middle-income countries are ranked low. These observations correspond with those by Misuraca (2007, ch.1 sec.1.3); that is, so far, governments are mainly concerned with e-govt rather than e-governance. The twin issues of information access and active citizen participation are challenges for governments (Saidi and Yared, n.d., IV.3.3), including through creation of an environment supportive of other e-gov players.

Implementation of E-Gov with Possible Impact on Democracy

Saidi and Yared (n.d. III.3) identify four broad areas of e-gov—administration, economic, social, and political—each with various activities (G2B, G2G, G2C, etc.), impacts, and examples of good practice. The political aspect alone (laws and regulations, decision-making process, strategies and policies, leadership) has various activities (e-participation in the form of concerted action, connecting government to citizens, joined-up government in terms of decision-making), assumed impacts (democratization reforms, strengthen accountability, speed up decision-making, improve quality of decision-making, enable innovative approach to government, increase transparency, create empowerment, strengthen capacity to investigate, develop and implement strategy and policy, anti-corruption drive), and examples (such as supporting free and fair elections in South Africa) (Saidi and Yared, n.d., II.3). In most of Africa, election is the most visible manifestation of democratic practice, so it makes sense to examine attempts to deploy ICTs in these realms.

Some African electoral authorities have deployed ICTs in voter registration and e-voting. With regard to registration, optical mark recognition

(OMR) and optical character recognition (OCR) have been used to record voter data in parts of Africa, such as in Tanzania's use of OMR (Pran and Merloe 2007, 46). The challenges of e-voting,[8] compared to e-registration, "appear much greater and certainly more controversial" (Joint 2004, 391). Although it is perceived to "excite new opportunities" and tackle electorate apathy (Joint 2004: 391), issues surrounding e-voting include: security (tampering with recorded votes and impersonation); secrecy (undue influence or traceability of voter identity); transparency (open to relevant policing authorities); accuracy (recording process); ease of voting (convenience and access); speed and efficiency (expedites or delays the process); effect on turnout (increases generally rather than benefit any particular social group); cost (justification) (Joint 2004: 393; Pran and Merloe 2007). Citing, among others, a 2001 Southern African Development Community report, Norms and Standards for Elections in the SADC Region, Pran and Merloe (2007, 54) point out that suspicion of e-voting is not surprising.

> Because of difficulties with the observation of the electronic voting, it is likely that society will be skeptical toward e-voting systems in any country and particularly where there is not an established record of holding elections in accordance with minimum international standards.

One case is that of Nigeria, with at least 60 million registered voters and 120,000 polling stations (Umonbong 2006). The Electronic Voting Machine (EVM), proposed as the fourth[9] component of the Electronic Voting System (EVS), was dropped for the 2007 general election after parliament rejected it (INEC n.d.). Critics cited unreliability, given the frequent power outage, although, to borrow from Claude Ake, more to blame would be the perceptions of "election malfeasance, as a variant of public corruption, symptomatic of the country's defective political culture" (Okafor 2006). Critical media reports at one point provoked a clumsy Independent National Electoral Commission (INEC) response (INEC, n.d.).

Ironically, Nigeria's migration from its legacy of paper-based voting system to EVS was "viewed as a means of ensuring free and fair elections" (ESRI 2007)—by the U.S.-based commercial vendor ESRI and the purchaser, INEC. The latter's argument was that "given the culture of election violence including ballot snatching, impersonation, ballot stuffing and vote rigging," EVS "will reduce to the barest minimum these unwholesome electoral malpractices" (Umonbong 2006). No one knows whether EVM would have made a difference, or whether the INEC was vindicated. Although the 2007 general elections resulted in a landmark power handover from one civilian government to another, the elections were condemned by monitors "for pervasive vote-rigging, violence, intimidation, and fraud...perpetrated by and with the connivance of the nation's security forces" (Okafor, blurb). To some observers, the rejection of EVMs was machinated by "anti-democratic forces among the political elite who could not change from their nefarious habit of old" (Agu 2008). All the same INEC implemented the other aspects of EVS that did not involve actual voting (ESRI 2007), but with much difficulty, and resulting in legal challenges (Pran and Merloe 2007, 40).

The Nigerian experience shows the consequences of assumptions about desirability of e-gov solutions, which lead to poor implementation strategies, especially inadequate preparation of public psyche and of the electoral agency. Based on seven dimensions for analyzing the success or failure of e-government, Boateng and Heeks (2003) examined an e-democracy application introduced in an unnamed West African country with the intention of making the electoral process more transparent (hopefully reducing post-election violence) and quicker. "Inscribed within the application's design were a number of inherent assumptions or requirements" (Heeks 2004, 4; Boateng and Heeks 2003)—about information, technology, processes, objectives/values, staffing/skills, management systems/structures, and other resources—incongruent with the implementation environment.

> Many of these elements will be what [Bruno] Latour refers to as prescriptions: requirements, or assumptions, or expectations about the context of the user of the e-government application...drawn from the world of the designer...there are dangers of a mismatch...(Heeks 2004, 6).

Perhaps this explains why e-election aspects of e-gov implementation in SA appeared to be less chaotic.

With at least 18 million voters and 15,000 polling stations in 400 constituencies (Mutula 2002), ICTs could help SA enhance efficiency and possibly transparency and accountability. In examining e-voting in the United States, the Netherlands, India, and Nigeria with a view to drawing lessons for SA, Masuku (2006, 108) argues that "best practices regarding the planning of an e-voting system in South Africa may yield important lessons of experience for other African countries." Pran and Merloe (2007, 21) point out that the Internet has been used in SA "to communicate polling stations assignments to voters." Under the theme, Connecting e-Citizens, Heeks (2002, 6) identifies the use, by SA's Independent Electoral Commission (IEC), of ICTs to support free and fair elections following the difficulties in the 1994 polls. IEC made it possible for members of the public to access the IEC Web site to verify their voting status and find out, among others, where they must vote (Theunissen 2001, 205).

> The effort included the creation of a nationwide satellite-based wide-area network and infrastructure; a bar-code system used to register

> 18.4 million voters in just nine days; a geographic information system used to create voting districts; a national common voters' role; a sophisticated election results centre for managing the process; and the training of 300,000 people. The massive program was completed in less than two years, in time for the vote (Heeks 2002, 6).

Similar details of the effort that won the IEC a major award in 2000 are provided by Mutula (2002), who points out that heavy-duty servers at a large call-center with fax facilities, and ability to collect and display results to the public, enabled even rural dwellers to participate in the electoral process.

> ICT was used for voter registration, the polling process, relaying of ballot, collection and verification, and relaying of results of the elections throughout the country... The application was largely successful as the electoral process was expeditious, long queues during voting were not experienced, and the electoral process was accepted by the great majority of stakeholders as transparent, free and fair. The results of the election were released in record time and all eligible voters were registered and able to vote. Communication was maintained between the election monitoring centre and the entire country (Mutula 2002).

Compare this with the disputed Kenyan 2007 general election, especially the presidential one, before which the Electoral Commission of Kenya (ECK) had rejected a mapping system among other ICT applications that would have enhanced the relay of accurate results. At an official enquiry into those elections, ECK chairman Mr. Samuel Kivuitu admitted in August 2008 that the exercise was riddled with manual-paper errors. Officially, Kenya's e-gov position includes "enhancing the provision of election services online to ensure that there is no congestion at polling halls and that vote counting is done quickly" (Okong'o 2005). The problem lies in implementation, as has been the case not only in Nigeria but also in Uganda.

In Uganda, a $22 million project to take digital photos of voters and match the images with the voters at polling stations in the 2001 general election was aborted due to the lack of political will; a "big bang" approach rather than a gradual introduction of the service; lack of capacity; and failure to involve the civil society (Anon 2002).

One aspect of implementation that needs to be taken into account is the fact that because most African ICT users fall in the grey area between groups that own ICTs and those that borrow access to ICTs from friends and relatives as well as group-access centers, there is a need for "re-intermediation models that insert a human intermediary between the citizen and the growing digital infrastructure of e-government" (Heeks 2002, 7). Such "intelligent intermediaries" could include professionals, public servants, NGOs and CBOs, or other public institutions (Heeks 2002, 7). It is for this reason that in the Golagang ("come together") public-private partnership, SA wanted to give the 1.3 million public service employees an opportunity to acquire Internet-connected PCs, although the project failed due to stakeholder fears of economic risks (Levin 2002).

Considered as least labor-intensive, a central government web portal is the common e-gov solution adopted by most African governments. In SA, the primary government web resource is the home page, www.gov.za, "where access to all current and recent legislation, bills, speeches, government documents, etc. is provided" although "provision is not, however, made for interactive services on these web pages" (Theunissen 2001, 204). The UN notes that there are some good pages.

> South Africa has a strong online presence. The website of the Department of Labor (www.labour.gov.za) in particular, is an excellent example of a public agency website that is well tailored to the needs of its stakeholders...a full-featured site that is a one-stop shop for labor issues (UN-DESA 2008, 26; Perry 2008).

The Web site of a similar ministry in Central Africa has also been similarly recognized. The Ministries of Labor and Social Welfare (www.mapess.gv.ao) of Angola "received high marks (80%)...a one-stop shop website" (UN-DESA 2008, 23). Back to southern Africa, the Ministry of Finance of Lesotho (www.finance.gov.ls) "permits its citizens to download forms...and also offers an online feedback mechanism that allows citizens to ask questions or make a suggestion" (UN-DESA 2008, 26). In North Africa, the Ministry of Education of Egypt (http://knowledge.moe.gov.eg/arabic/) "has improved its website by making it more interactive" while the Ministry of Finance of Morocco (www.finances.gov.ma) "allows its citizens to create accounts online, download financial statistics and retrieve archived information..." (UN-DESA 2008, 25).

West African countries such as Cameroon started creating portals as far back as 2001 (Olivier 2002). One portal has been singled out for its e-participation excellence. The Burkinabe national portal (www.primature.gov.bf) "is the only African portal which allows for online consultation" (UN-DESA 2008, 27). Other enhanced interactive Web sites in the region are those of the Ministry of Finance of Cape Verde (http://www.minfin.cv) and the Ministry of Health of Senegal (www.sante.gouv.sn). Nigeria Direct (www.nigeria.gov.ng) official information gateway features opinion polls, with questions such as: "Should public office holders enjoy immunity?"[10] It also has provision for feedbacks, a helpdesk, and searchable directory, besides an interactive map with all states clickable—enabling access to basic information about each state and a link to their web portal. An interesting feature is "How do I?" link in various fields.

Eastern Africa boasts an e-gov pioneer in Rwanda (sometimes grouped in Central Africa), even though its E-Rwanda Project emphasizes mainly service delivery, and the portal, www.rwandagateway.org, is mostly information-based rather than interactive. Their Ministry of Finance Web site (www.minecofin.gov.rw), available in English and French, has been "improved" (UN-DESA 2008, 24). In the same region, the Ministry of Education of Mauritius (http://ministry-education.gov.mu) "allows citizens to register online and download forms..." (UN-DESA 2008, 24). Also singled out for interactivity is Kenya's portal.

> The welcome page (www.kenya.go.ke) is well laid out and easy to navigate...a step towards a "one-stop shop" design...Kenya's example shows how even countries with constrained resources can make solid progress in e-government (UN-DESA 2008, 24).

The last sentence about progress in e-government is in reference to the specialist directorate referred to above. The Kenyan portal has provision for users to register and log in as well as to see "who's online," including guest (non-registered) users. There is provision for downloads (e.g., police assault or property-loss reporting forms) and links to information about constituencies, citizens, and parliament, among others. The parliament link has a "Parliament Tracker" section, for monitoring bills, to order papers, legislative calendar, and committee businesses, as well as provisions for downloads, virtual tour, links to ministries and display of date and time. The e-government pages provides information on "the Digital Village Network (DVN)...a mega-community initiative coordinated by ICTvillage.com that brings together individuals and organizations in the Government, public, private and civil society sector under one umbrella to radically transform Kenya's economy, society and politics using ICT."[11] The e-Citizenship link provides, among others, information about "who can vote" under Voting and Elections. There is a Discussion Forum, but with hardly any active discussions; there is also no clear provision for contact or feedback.

Kenyan legislators were perhaps spurred on to support e-gov after some of them visited Rwanda and remarked: "The system of governance and technology in Rwanda makes our Parliament look Victorian and archaic" (NT 2005). In 2008, newly elected Speaker of the Kenyan National Assembly, Mr Kenneth Marende, in his acceptance speech, said the voting process in Parliament would have been less tedious had it been conducted electronically (Limo 2008). Yet MPs in the previous parliament exhibited their conservative streaks when some of them objected to their profiles appearing on the parliament Web site. This reinforces the findings of Taylor and Burt (2001, 34–36), that a number of parliament web pages "have only basic information about the institution and key leaders (such as MPs) but hardly provide citizen services or provide for active citizenship or support electronic access or have innovative features." They elaborate:

> Far from this [democratic] deficit [in many Western political institutions], Parliaments, as expressed through their websites, are locked in to a parliamentary model of democracy that prevents a more informative, expressive, political, knowledge developing and activist approach to relationships with citizens (Taylor and Burt 2001, 38).

For Kenyans, even if Parliament is not fully reformed, "e-governance or digital democracy will be the guiding principle in the way government products and services are provided to the people" (Okong'o 2005). The government promises to "introduce and enhance e-talking to citizens by providing citizens with government publications…through websites" (Okong'o 2005). Other initiatives in the e-strategy policy include "enhancing listening to citizens by increasing the input of citizens into public sector decisions and actions" (Okong'o 2005). However, it will take major campaigns, such as the August 2008 "Public Service Week," and culture change for the mindset of public servants, often insistent on face-to-face interactions, to change.

As more and more African governments go online at the national level, with the exception of a few "laggard" countries such as CAR, Somalia, and Zambia (UN-DESA 2008, 46), some are already going online at the local levels. One example is Gauteng Online (www.gautengonline.gov.za). "Once the plan is fully operational, citizens will be able to interact with government at any time with minimal effort" (GPG n.d.). Members of the public without Internet access can phone the Gauteng Provincial Government (GPG) contact centre to access the information on the portal. Other access channels include "multi-purpose community centers, digital villages, Gauteng Online computer laboratories in public schools and remote mobile computing vehicles" (GPG n.d.). In this way, basic e-government equations—G2C, G2B, G2G—get exponentially reproduced at the local level, LG2C, LG2B, and LG2Gn (where n = levels of government in a specific institutional setting) (Misuraca 2007, ch.4 sec.4.1).

Conclusion

E-gov is not a panacea to problems of governance and democracy. In fact, some examples in this chapter show that e-gov could complicate matters, especially if poorly and hurriedly implemented. Problems at the U.S. presidential elections of 2000 and several cases of loss of sensitive personal data in the UK show that there are e-gov hiccups even in the most advanced industrial nations.

Echoing Coleman (in UN-DESA 2008, 121), Heeks (2002), and Misuraca (2007), I would argue that homegrown solutions need to be seriously considered for implementing e-gov in Africa, with local content in mind and a possible adoption of flexible regional or pan-African strategy suggested by Oyawoye (n.d.)—but bearing in mind that no single strategy fits all (Misuraca 2007). Already, the NEPAD e-Commission has been attempting to undertake work in such a direction. A second aspect is the need to strengthen Public-Private Partnerships, attempted in e-gov in places such as Singapore, to provide not only investment but also enrich the pool of what Heeks (2002) terms "intelligent" intermediaries so necessary for e-gov to succeed, especially in the initial stages, in Africa. A third imperative is the need to constantly evaluate the political participation, transparency, and accountability aspects of e-gov initiatives, with a view to improvement rather than abandonment. The fourth aspect is the need to seriously exploit the "m-opportunity," given the increasing number of cell phone subscribers-users in Africa, with projections of over 300 million subscribers by 2009 (Mendes et al. 2007, citing Joss Gillet, 2007 in Wireless Intelligence). This offers an excellent, and rapidly growing, base from which to build access not only to banking but also other services through simple cell phone applications (Mendes et al. 2007, 40, 42).

Finally, to inspire public confidence, acceptance, and trust, Africa needs to learn from its own mistakes, as well as those of others, and take into consideration issues of data security. This is more crucial in Africa, where there is less experience of, and enthusiasm for, interaction with government through technology than elsewhere.

Notes

1. See for instance Norris 2007.
2. For more details, see the relevant World Bank web links at www.worldbank.org/egov
3. See also Misuraca 2007, ch. 3 sec. 3.2; Chissick and Harrington 2004, 6–10.
4. See source for the other three.
5. It is these parameters that Damn and Thomas (2006) use to privilege a closer look at China's cyberspace politics.
6. This is part of the Computerized Nigeia Project. For details, visit the Web site of Zinox Technologies: http://www.zinoxtechnologies.com/newspl.asp (accessed December 20, 2008).

7. The discourse of "immense power" of ICTs, "empowering" millions, is also highlighted by Theunissen (2001, 203).
8. Types: telephone; SMS via mobile phones; Internet, via PCs; electronic voting machines in a variety of places; digital TV; ATM or other bank machines (Joint 2004: 393–394).
9. "Components of EVS are: Electronic Voters Register which had been in operation since 2002 and was used for compiling the voters register used in 2003 elections. The second component is Electronic Authentication, while the third is the Speedy Transmission of election results" (INEC, n.d.).
10. Gleaned during a visit in August 2008.
11. See http://www.e-government.go.ke/index.php?option=com_content&task=view&id=57&Itemid=1.

References

Agu, E. 2008. "Maurice Iwu—The Chicken is Coming Home to Roost." *Daily Champion*, July 2, 2008. http://allafrica.com/stories/200807020374.html (accessed July 5, 2008).

Anon. 2002. Failed Electronic Voter Registration in Uganda. Success/Failure Case Study No. 22, eGov4Dev Information Exchange. Manchester: IDPM, University of Manchester. http://www.egov4dev.org/success/case/iecuganda.shtml.

Bellamy, C. and J. A. Taylor. 2003. *Governing in the Information Age*. Maidenhead and Philadelphia: Open University Press.

Boateng, K. and R. Heeks. 2003. Computerizing Election Results Management in West Africa, Real-World Design-Reality Example 2, eGov4Dev Information Exchange. Manchester: IDPM, University of Manchester. http://www.egov4dev.org/success/case/electcomput.shtml (accessed July 3, 2008).

Chissick, M. and J. Harrington. 2004. *E-government: A Practical Guide to the Legal Issues*. London: Sweet and Maxwell.

Damn, J. and S. Thomas. 2006 *Chinese Cyberspaces: Technological Changes and Political Effects*. Abingdon: Routledge.

Dode, R. O. 2007. Prospects of e-Government Implementation in Nigeria. Proceedings of the First International Conference on Theory and Practice of E-Governance, Macao, China. ACM International Conference Proceedings Series Vol. 232: 380–383. New York: ACM. Available at http://portal.acm.org/citation.cfm?id=1328057.1328137 (accessed July 3, 2008).

ESRI. 2007. Nigerian Voting System Modernized: GIS Prevails Despite Unnavigable Terrain and Political Unrest. Washington, DC: ESRI. http://www.esri.com/news/arcnews/summer07articles/nigerian-voting.html (accessed June 28, 2008).

GPG. n.d. Gauteng Online: Government at the Click of a Button. Gauteng Provincial Government.

Heeks, R. 2002. *e-Government in Africa: Promise and Practice.* No.13, iGovernment Working Papers Series. Manchester: Institute of Development Policy and Management. http://www.sed.manchester.ac.uk/idpm/research/publications/wp/igovernment/documents/igov_wp13.pdf (accessed July 4, 2008).

Heeks, R. 2004. *e-Government as a Carrier of Context.* No.15, iGovernment Working Papers Series. Manchester: Institute for Development Policy and Management. http://www.sed.manchester.ac.uk/idpm/research/publications/wp/igovernment/documents/igov_wp15.pdf (accessed July 4, 2008).

INEC. n.d. Misleading Editorials on the Use of EVS. http://www.inecnigeria.org/newsview.php?news=15&newsid=116 (accessed December 20, 2008).

Joint, A. 2004. "Electronic Voting." In *eGovernment: A Practical Guide to the Legal Issues*, ed. M. Chissick and J. Harrington, 391–411. London: Sweet and Maxwell.

Kalu, K. N. 2007. "Capacity Building and IT Diffusion: A Comparative Assessment of E-Government Environment in Africa." *Social Science Computer Review* 25(3): 358–371, http://ssc.sagepub.com/cgi/content/abstract/25/3/358 (accessed July 6, 2008).

Krishna, S. and S. Madon, eds. 2003. *The Digital Challenge: Information Technology in the Development Context.* Aldershot: Ashgate (also available as Google e-book).

Levin, A. 2002. (Not) Providing Computers for All South African Civil Servants. Success/Failure Case No. 26. eGov4Dev Information Exchange. Manchester: IDPM, University of Manchester. http://www.egov4dev.org/success/case/golaganang.shtml (accessed July 3, 2008).

Limo, A. 2008. "Electronic Voting Will Greatly Enhance Democracy." *Daily Nation*, January 18, 2008, Opinion section. http://allafrica.com/stories/200801180801.html (accessed July 4, 2008).

Masuku, W. K. 2006. An Exploratory Study on the Planning and Design of a Future E-Voting System for South Africa. Mphil Thesis. South Africa: University of Western Cape. http://etd.uwc.ac.za/index.php?module=etd&action=viewtitle&id=gen8Srv25Nme4_7837_1183351619&allowManage= (accessed July 4, 2008).

Mendes, S., E. Alampay, E. Soriano, and C. Soriano. 2007. The Innovative Use of Mobile Applications in the Philippines—Lessons for Africa, SIDA. http://siteresources.worldbank.org/EXTEDEVELOPMENT/Resources/20071129-Mobiles_PH_Lessons_for_Africa.pdf?resourceurlname=20071129-Mobiles_PH_Lessons_for_Africa.pdf (accessed July 5, 2008).

Misuraca, G. 2007. *E-Governance in Africa: From Theory to Action—A Handbook on ICTs and Local Governance.* Africa World Press/IDRC. http://www.idrc.ca/en/ev-113398-201-1-DO_TOPIC.html (accessed July 2, 2008).

Mutula, S. M. 2002. Supporting Democracy with ICTs: South Africa's Independent Electoral Commission. eGov4Dev Information Exchange. Manchester: IDPM, University of Manchester. http://www.egov4dev.org/success/case/iec.shtml (accessed July 3, 2008).

Narayan, G. 2007. "Addressing the Digital Divide: E-Governance and M-Governance in a Hub and Spoke Model." *The Electronic Journal on Information Systems in Developing Countries* 31(1): 1–14.

NT. 2005. "Kenyan MPs Humbled by ICT in Parliament." *New Times*, September 12, 2005. http://www.rwandagateway.org/article.php3?id_article=973 (accessed July 3, 2008).

Okafor, V. O. 2006. Nigeria's Electoral Challenge: What is to Blame, the Ballot Box or the Political Culture? http://www.utexas.edu/conferences/africa/ads/1160.html (accessed June 28, 2008).

Okafor, V. O. 2008. *Nigeria's Stumbling Democracy and Its Implications for Africa's Democratic Movement.* United States: Praeger Security International.

Okong'o, V. 2005. "The E-Government Experience in Kenya: The Story So Far." In *At the Crossroads: ICT Policymaking in East Africa*, ed. F. Etta and L. Elder, doc. 17 of 26. East African Education Publishers / IDRC. http://www.idrc.ca/en/ev-93058-201-1-DO_TOPIC.html (accessed July 2, 2008).

Okpaku, J. O. 2007. Foreword to *E-Governance in Africa: From Theory to Action—A Handbook on ICTs for Local Governance*, by G. Misuraca. Africa World Press/IDRC. http://www.idrc.ca/en/ev-115656-201-1-DO_TOPIC.html (accessed July 2, 2008).

Olivier, K. T. 2002. The Cameroon Government Web Portal. eGov4Dev Information Exchange. Manchester: IDPM, University of Manchester. http://www.egov4dev.org/success/case/camportal.shtml (accessed July 3, 2008).

Osimo, D. and Y. Punie. 2008. "The Emergence of Social Computing: Implications for Government." *Intermedia* 36(2): 26–31.

Oyawoye, T.D. n.d. An African Perspective in Implementing E-Governance. www.ictingovernment.co.za (accessed June 23, 2008).

Parkinson, S. 2005. *Telecentres, Access and Development: Experience and Lessons from Uganda and South Africa.* Practical Action Publishing / Fountain / IDRC. http://www.idrc.ca/en/ev-87255-201-1-DO_TOPIC.html (June 23, 2008).

Perry, S. 2008. E-Governance in Africa Goes Backward. ITWEB. http://www.itweb.co.za/sections/computing/2008/0801081044.asp?A=AFNandS=All%20Africa%20NewsandO=F (accessed June 23, 2008).

Pran, V. and P. Merloe. 2007. *Monitoring Electronic Technologies in Electoral Processes: An NDI Guide for Political Parties and Civic Organizations.* Washington, DC: National Democratic Institute for International Affairs.

Prins, C. 2001. "Electronic Government: Variations on a Concept." In *Designing E-Government: On the Crossroads of Technological Innovation and Institutional Change*, ed. J. E. J. Prins, 1–5. London: Kluwer.

Saidi, N. and H. Yared. n.d. E-Government: Technology for Good Governance, Development and Democracy in the MENA Countries. Washington, DC: The World Bank. http://www.worldbank.org/mdf/mdf4/papers/saidi-yared.pdf (accessed June 23, 2008).

Taylor, J. A. and E. Burt. 2001. "Pluralizing Teledemocracy: Not-for-Profits in the Democracy Polity." In *Designing E-Government*, ed. J. E. J. Prins, 29–39. London: Kluwer.

Theunissen, C. A. 2001. "Applications and Implications of Cellular (Mobile) Telephone Technologies for Electronic Government and information: An Overview." In *Designing E-Government*, ed. J. E. J. Prins, 197–212. London: Kluwer.

Umonbong, O. 2006. A Paper on the Voting System in Nigeria. Presented at the AEA Seminar, Blackpool, UK. http://www.aea-elections.co.uk/annualseminars/downloads/nigeria_text.pdf (accessed June 28, 2008).

UN-DESA. 2005. *Global E-Government Readiness Report 2005: From E-Government to E-Inclusion*. New York: United Nations Department of Economic and Social Affairs. http://www2.unpan.org/egovkb/global_reports/05report.htm (accessed July 23, 2008).

UN-DESA. 2008. *UN E-Government Survey 2008: From E-Government to Connected Governance*. New York: United Nations Department of Economic and Social Affairs. http://www2.unpan.org/egovkb/global_reports/08report.htm (accessed July 23, 2008).

White, G. 2003. Creating an Enabling Environment for Accessing Government Services. Presentation to the SMS Conference, Towards an Integrated Public Service, September 16, 2003.

World Bank. 2008. www.worldbank.org/egov.

6, P. 2001. "E-Governance: Do Digital Aids Make a Difference in Policy Making?" In *Designing E-Government*, J. E. J. Prins, 7–27. London: Kluwer.

CHAPTER 3

"Misclick" on Democracy: New Media Use by Key Political Parties in Kenya's Disputed December 2007 Presidential Election

George Nyabuga and Okoth Fred Mudhai

While the use of "new" media by key political parties and presidential candidates has intensified in recent Kenyan presidential elections, the controversial December 2007 poll laid bare the limits of technology's role in democracy. We argue that while new media may have some potential to help monitor and mobilize political activity, and possibly encourage political engagement, they can also reinforce the positions of those in power not only due to their limitations but also by their manipulability by scheming human agents.

Introduction

On August 21, 2007, the then ruling party of Kenya, the National Rainbow Coalition (NARC), announced it was going high-tech in preparation for election duel, later set for December 27. President Mwai Kibaki's[1] campaign team in August 2007 ostensibly "unveiled a hi-tech communications network that will take the use of short text messages and satellite telephones in electioneering to another level."[2] The "hi-tech" team would use GSM cell phones and satellite telephony to provide faster communication and offer effective coordination between teams based in rural and remote areas, with a command centre in the capital,

Nairobi. This was not, however, surprising. Various political parties and politicians had years earlier integrated information communication technologies (ICTs) into their communication systems. Political parties such as the NARC and Kibaki's former party, Democratic Party of Kenya (DP), and presidential aspirants such as Raila Odinga, Kibaki, Uhuru Kenyatta, and Simeon Nyachae used the Internet in the 2002 general elections. So while the announcement by the Kibaki team did not necessarily mark a significant shift in adoption of new media technology, it demonstrated the powerful allure these media have over political elites in Kenya. Such facilities undoubtedly provided, and still offer, Kenya's political elite the opportunities to directly seek support and influence or "manipulate" political behavior.

In this regard, this chapter argues that while new media can mobilize for political activity and encourage political engagement, it can also reinforce the positions of the political elite or those in power (cf. Norris 2001). Mobilization theorists are of the view that new media are capable of encouraging and supporting political activism, whereas reinforcement theorists hold that new media can only strengthen but not radically transform participatory politics (Norris 2001, 2002). Accordingly, the growing use of the Internet and cellular telephony is seen as both empowering some Kenyans to actively engage with the leadership, the political parties, and political systems, and at the same time reinforcing the status quo by offering greater leverage to those already in power, or to the powerful political elite (cf. Norris 2001, 2002). Yet such analysis may seem simplistic given that the role and influence of new media depend on a variety of uses different political actors put these facilities to. However, the mobilization and reinforcement (Norris 2001, 2002) perspectives, or what Ward, Gibson, and Nixon (2003) call equalization and normalization approaches, seem appropriate frameworks for analyzing both modalities and effects of new media use in Kenya's 2007 presidential election.

This chapter concentrates on the uses and possible consequences new media have on the political process, with a specific focus on Kenya's presidential election, mainly with regard to the Internet but also considering other platforms, particularly cellular telephony and short messaging service (SMS or texts). In charting the use of new media in electoral politics in Kenya, this chapter addresses various questions, among them: how do political actors use new media? Is there any significant change from the way they use the new media from their utilization of the "traditional" forms of media? These questions, among others, are critical to undertake research into the use of cellular telephony and Internet in current campaigns, given the fact that new media "refashion" or "remediate" older forms or "traditional" media (Bolter and Grusin 1999).

The chapter starts with a brief look at Kenya's political history. It then examines the development and diffusion of the Internet and cellular telephony in Kenya before proceeding to examine their use in the 2007 presidential election.

The Legacy of Kenya's Postcolonial Political History on Current New Media Usage

Kenya can be cautiously called a "democracy." This caution is premised upon recent events, particularly glaring electoral anomalies in the December 2007 election and subsequent violence that plagued the country following the announcement of the results, events that dented the country's "democratic" and political reputation and battered the achievements since the country became a multiparty state in 1991.

On attaining independence on December 12, 1963 after seventy-five years of British rule, Kenya adopted a Westminster constitutional model providing for political pluralism, but multipartyism lasted only a year. After numerous defections to the then ruling party, Kenya African National Union (KANU), the only opposition party at the time, the Kenya African Democratic Union (KADU) "voluntarily" disbanded, making Kenya a de facto single-party state. In a construction of what Thomson (2004, 111) calls "consensus" politics, the decimation of the adversarial multiparty political systems was considered necessary as newly independent African countries attempted to stave off the "divisive and perilous" competition in order to "save" the fragile African states (Nyerere 1968; Osaghae 2003; Thomson 2004). KADU disbandment gave Jomo Kenyatta the opportunity to consolidate and personalize power, marking the start of his serious clampdown on political opposition and criticism until his death in 1978, even though Throup and Hornsby (1998, 15) believe that political life was "remarkably open...free by African standards." Following a failed military coup, Daniel Moi's regime changed the constitution in 1982 to make Kenya a de jure one-party state. During this period, Moi effectively "destroyed all possible centers of alternative powers...thus ushering in an authoritarian and oppressive state that virtually turned Kenya into a police state" (Kibwana, Wanjala, and Owiti 1996, 118). The formation of political parties became illegal and those who attempted to challenge Moi's and KANU's position were detained without trial, jailed after numerous sham trials, or forced underground or into exile.[3] Not even ministers and government officials were spared. The legislative process became ineffective as Parliament turned into a "mere rubber-stamp" for Moi's wishes and demands (Throup and Hornsby 1998; Adar and Munyae 2001; Kanyinga 2003; Odhiambo-Mbai 2003). Mainstream media and communication channels were reduced to government mouthpieces as free expression was curtailed, forcing political actors to resort to *mwakenya*, underground means of exchanging political views—such as clandestinely printed and distributed leaflets.

The reintroduction of multiparty politics in 1991 engendered some changes to the political process, expanding the political space but not radically altering the power structures to allow for a responsible and accountable leadership (cf. Throup and Hornsby 1998; Adar and Munyae 2001; Oyugi 2003). Nevertheless, besides permitting the formation of political parties and subsequent competition for power against KANU, multipartyism allowed and guaranteed Kenyans the right to free speech, assembly, and the right to contend for political power from 1992. Yet, the political environment remained heavily skewed in favor of KANU, as the political establishment and elite were not responsive and adoptive of the "new" and changing political climate (Throup and Hornsby 1998; Adar and Munyae 2001; Oyugi 2003). Foremost among the problems experienced was lack of access to channels or media of mass communication, considered vital to effective political communication, inclusion, and participation.

In essence, the media of mass communication in Kenya were for many years under state control. This allowed the government and KANU to manipulate or dominate the use of state-owned media. Unsurprisingly, therefore, they jealously guarded public, particularly broadcast, media against encroachment from recalcitrant political actors, opposition politicians, parties, and organizations seeking liberalization as a precursor to political pluralism (Ochieng 1992; Ochilo 1993; Throup and Hornsby 1998; Kibara 2003). The state-owned Kenya Broadcasting Corporation (KBC) was often dominated by KANU; news especially about President Moi was accorded 80 percent positive or favorable coverage, while the opposition only received negative reportage (Kibara 2003, 287). In 1992, while "99 percent of KANU's activities were covered positively and only 1 percent negatively by KBC radio and TV news... only 4 percent of opposition activities were covered positively" (ED, CJPC, NCCK 1998, 95). This disparity was the basis for the opposition's incessant complaints of media bias (Throup and Hornsby 1998, 365), upheld by the Commonwealth Observer Group, (1993, 27) which confirmed that most media, and particularly KBC, "devoted its news coverage to chronicling the comings and goings of the President [Moi], his Ministers and various government functionaries."

Paradoxically, this again became the norm—though not to the extent of the Moi era—in Kibaki's administration, regardless of promises that political parties would be accorded fair and equal reportage, especially in state-controlled and public-funded media. Surveys around 2007 elections indicated the incumbents enjoyed more and better coverage in the mainstream media, which was often intimidated into self-censorship. Indeed, in early 2006, the newspaper and TV premises of The Standard Media Group were raided by hooded state security agents who destroyed or carried away equipment and brutalized the staff to stifle their reporting on a sensitive political matter. In effect, doubts still remain over the commitment of the leadership to democracy and to democratic ideals, such as free expression and freedom of information (cf. Barkan 2004, 2007; Murunga and Nasong'o 2006, 2007; Cooke 2007). Given such a situation, new media has become an attractive, though alternative, channel of communication for a variety of political actors including politicians (both in and outside government) and parties seeking to convey their messages or equalize the political playing field (cf. Ward, Gibson, and Nixon 2003).

New Media as Agents of Change

Development, Diffusion, and Use of New Media in Kenya

There has been a dramatic growth of the use of the new media in many parts of Africa, and the continent is said to have the highest growth rates of mobiles and the Internet, but the growth rate indicators and real use are incongruent. Even though International Telecommunication Union (ITU)'s recent indicators show Africa's cellular phone growth rate is nearly twice as high as Europe's and Asia's, in real numbers Africa falls way back in global rankings of actual use.[4]

Statistics indicate Kenya has one of the highest number of Internet and cellular telephone users in sub-Saharan Africa. In 2008, there were about 15 million cell phone subscribers (and perhaps many more users) in Kenya, from a low of only 15,000 in 1999, a dramatic growth in about a decade.[5] By comparison, statistics indicate that there were in 2008 about three million Internet users in Kenya, even though in reality this number is most likely to be higher due to the proliferation of cybercafés, telecenters, and public places offering Internet services, as well as the introduction of cellular broadband by the Vodaphone-backed dominant GSM service provider, Safaricom. There were in 2004 over 1,000 cybercafés in Kenya from a low of about 100 in 2000 (GoK 2004a). The Internet is also accessible in 400 post office telecenters around the country, and this was expected to expand to all the 890 outlets of the Postal Corporation of Kenya by the end of 2007 (GoK 2004a, 39; GoK 2004b, 191–192), even though there were contractual hiccups. Such expansion would mean that even those in the remotest parts of the country may potentially access the Internet, especially because 80 percent of Kenyans live in the "marginalized" rural areas (GoK 2004a). The number of Internet users is especially anticipated to grow substantially if a $135 million World Bank funded Rural Digital Village "Marshall Plan" unveiled in May 2007 is completed successfully. Although the completion date is not clear, the project seeks to connect the whole country in what is seen as the creation of Internet villages. Once complete, the initiative would undoubtedly provide easier access to the Internet even in the remotest parts of the country.[6]

The above examples reveal the adoption differentials between PC-based Internet and cellular telephony. While PC-based Internet has been available since the mid-1990s, cellular telephony did not take off until about 2001 when the cost of handsets and connection began to fall due to subsector liberalization. Moreover, whereas PC-based Internet requires significant resources for infrastructure development, hardware such as computers and modems, software, and skill acquisition to ensure computer and web literacy, cellular telephony infrastructure is comparatively cheaper, the cost of sets keep falling and, more importantly, they are relatively easy to use. This can be explained using the technology acceptance model (TAM) proposed by Davis in 1989. TAM is grounded on two important components, perceived usefulness and ease of use. The adoption rates can also be explained using the diffusion of innovation theory. Rogers (2003, 5) identifies five stages of diffusion: awareness, interest, evaluation, trial, and finally adoption. The adoption of technology is influenced by, among other factors, the relative advantage derived from the new technology. This explains the astounding pace of cellular phone adoption in the space of just about a decade in Kenya.

Given the availability of these resources, then, how did the presidential candidates use them? What possible effects did they have on their campaigns? Were they able to sway or influence public opinion by using cellular telephones and the Internet? The section below attempts to answer these questions.

The Use of New Media in Kenya's Presidential Elections

The idea that new media could reinvent and strengthen political parties (Stromer-Gally and Jamieson 2001; Ward, Gibson, and Nixon 2003; Buckler and Dolowitz 2005; Chadwick 2006) seems to have attracted parties and top politicians in Kenya. Indeed, several political parties including KANU, NARC, DP, and FORD-People went online prior to the 2002 elections.

The new media had two important advantages for political parties and candidates, particularly those in the opposition camps. First, they provided them with the means of getting their own message across without the interference, scrutiny, and questions from information gatekeepers, such as journalists and even government censors. Second, they allowed political actors greater control of their own content or messages (cf. Ward, Gibson and Nixon 2003; Buckler and Dolowitz 2005). These are the advantages enterprising politicians earmarked in 2005 after the constitutional referendum exposed political differences among Kenya's top political class. The period after the referendum saw heightened political activity ahead of the 2007 poll. As such, the contenders for Kenya's top job had begun lining themselves up, taking their campaigns to the country while others started touring various destinations around the world to lobby Kenyan expatriates and backers for support. In this case, modern forms of political communication came in handy, and the Internet became an attractive medium and platform for their political messages. Given the competitive nature of Kenya's current politics, the utilization of new media forms, particularly the cell phone and the Internet were a means of vote-seeking and mobilization of political activity, although their adoption by a largely elitist cohort of politicians was seen more as a reinforcement strategy.

The trend, however, predated the 2007 elections and can be traced to the late 1990s and early 2000s when some parties and political actors went online for the first time. For example, although initially opposed to the idea of media pluralism before the 1997 elections, KANU was the first party to embrace the Internet. "KANU has finally entered the 21st century, we are now in the digital age," announced the then party secretary general, Joseph Kamotho, during the launch of its Web site (www.kanu-kenya.org) in June 2001.[7] The speech by Moi, the then president and party leader, extolled the benefits of IT and the information superhighway.[8]

Even though KANU's online presence came to be regarded as a cynical, survivalist, and reinventionist strategy prior to the 2002 general election, it was a seminal development for the ruling party. Other parties followed suit in a highly competitive electoral environment. In trying to sell itself in the political market, KANU posted on its Web site its election manifesto detailing its strategies, plans, ideology, and achievements in power. Predictably, none of its litany of failures was on the site. Similarly the DP (www.dp-kenya.org) launched its Web site the same year. NARC (www.narc-kenya.org) and FORD-People (FORD [Forum for Restoration of Democracy]-P) launched their Web sites in 2002. Presidential aspirants Uhuru Kenyatta (www.uhurukenyatta.co.ke), Kibaki (www.kibakiforpresident.net), Simeon Nyachae (www.nyachae.org), and Joe Donde (www.joedonde.com) also launched their Web sites, providing information about themselves, their policies, and political agendas to capture a growing constituency of Kenyan Internet users (IED 2003; Otenyo 2003).

However, an examination of the Kenyan political parties' Web sites before the 2002 general election did not indicate any significant departure from their usual modi operandi. Most of the material online was about party personalities. For example, most of the information in the NARC and DP Web sites was about their candidate Kibaki. The sites also did not have any information on the purpose of elections, or the role of political parties, politicians, and government. Neither was there any facility for interactive communication. According to various interviewees, these online activities were seen not only as a way of communicating directly with the people but also capturing a constituency of younger voters or diaspora Kenyans who were more frequent users of the Internet than consumers of local traditional media and who hold significant sway in local politics. "The target was those who were out of reach of the media," said Ramadhan Juma, the DP National Executive Officer.[9] Although admitting that it had little overall impact on their campaigns due to the limited reach in rural Kenya where most of the voters live, their online presence was nonetheless vital to the growing Kenyan "middle-class and opinion shapers," according to Juma. "We did not have to worry about the editing of our press statements by journalists...," Juma pointed out. "We put out the information the way we wanted it and in this way eliminated the middlemen, the errand boys."[10]

Even though it is debatable whether the Internet eliminated the political middleperson, there is no denying that the prospect of unedited information or content sent directly to the electorate is attractive to various political actors in Kenya who oftentimes complain of biased or imbalanced reports. This is particularly important to opposition parties in the country, as they still receive little attention on state-controlled media, especially the most effective ones like radio and TV.

With the NARC government increasingly turning autocratic, and the political space increasingly suffocated by the "new" leadership (Murunga and Nasong'o 2006; Nyabuga 2006), the Internet and cellular telephony became major channels of political communication, particularly from opposition parties and politicians. In various electoral contests, well-known businessman-cum-politician John Harun Mwau is famed for his use of powerful radios and satellite phones during intense campaigns involving his party of choice. In effect, these media have enabled opposition politicians and parties to circumvent official censure to communicate directly with Kenyans. This aspect was captured on May 19, 2007 in an editorial cartoon in Kenya's leading newspaper, the *Daily Nation*, by the country's best-known cartoonist—Godfrey Mwampembwa, Gado. Titled, "NARC-K goes to elections..." the cartoon attempted to illustrate a miniature character within it dubbed "mobile democracy"—a hand holding a cell phone handset with telling words displayed on the screen:

> Dear Member, U can now vote 4 the candidate of ur choice by just sending an SMS with the name of ur preferred...

Such direct communication has not only allowed political parties but also individual information consumers to control their menu of available media and information choices. Essentially, this has enabled the individual information consumer to determine what to use or discard, depending on the uses and gratifications they hope to derive from their interaction. That Internet and cell phone users, like many other media consumers, are self-selecting—consuming new media products based on their preferences rather than mass appeal (Hill and Hughes 1998, 44)—reinforces the idea that new media in general and the web in particular "remediates" older media (Bolter and Grusin 1999). On this score, the Internet and cellular telephony in Kenya have not really changed the way people do things. Even if they did bring about any change—and this might take some time—they will mostly allow them to do the same things differently, as numerous scholars, for example, Margolis and Resnick (2000), Kamarck (2002), and Margolis, Resnick, and Levy (2003), posit.

Regardless, communication and political experts see the online presence of political parties and politicians as a move toward the reengagement of people in open political discourse aimed at enhancing participatory politics (cf. Ward, Gibson and Nixon 2003). This idea seems to have informed what many believe to be the web's capacity to open the political space in Kenya. "It is a good thing that political parties and other political organizations have started utilizing the Internet as a communication tool," says Moses Owuor of the International Foundation for Election Systems (IFES) in Kenya.[11] He considers these media as the key to an open or transparent political process due to their capacity to allow greater and open debate among citizens, and also between ordinary people and their leaders. "There is a lot of movement toward reform, to open up space for communication... the internet is a political resource that has the capacity to reform politics in this country."[12] It would thus seem, at least in the eyes of those who establish and use these sites in Kenya, that the Internet and cell phones are expanding and transforming the public or political space and offering a direct line of communication with citizens (cf. Grossman 1996; Norris 2001; Stromer-Galley and Jamieson 2001; Keren 2006; Stanyer 2007).

However, even though the Internet is sometimes considered a user-friendly and user-controlled, many-to-many medium (Grossman 1996; Loader 1998; Tsagarousianou, Tambini and Bryan 1998; Hacker and van Dijk 2000), current utilization of the web by the political leadership in Kenya indicates that there is almost a one-way communication process dominated by the political party and political leadership. This is because many of the political parties' or politicians' Web sites have neither interaction facilities nor allow feedback or correspondence. The sites are essentially sources of information that the politicians or parties hope will be accessed by citizens. The Web sites also serve as platforms through which the politicians and parties address people directly. Given this scenario, if its development is left only to the politicians, it would seem doubtful that the web can become a two-way media of interaction among the parties, the candidates, and the *mwananchi*, or the "ordinary" citizen/electorate. Even though this is somewhat changing as political actors develop interactive facilities on their Web sites, politicians and political leaders do not always respond to feedback or enquiries. Although this could be the result of volume and workload, what Postman (2004, 4) calls "information glut," it has become clear over time that they did not necessarily intend to enhance interactive communication as much as sending their messages directly to intended consumers.

The above observations are reinforced by events preceding the December 2007 poll. As preparations for the 2007 elections advanced, political parties and politicians started making a comeback onto the Internet after abandoning their Web sites immediately after the 2002 polls. Odinga and his new party, Orange Democratic Movement, were among the first to develop a Web site (www.raila2007.com). Kalonzo Musyoka (www.kalonzomusyokaforpresident.co.ke), William Ruto (www.williamruto.com), Musalia Mudavadi (www.mudavadiforkenya.com), Najib Balala (www.najibbalala.net), Julia Ojiambo (www.electjulia.com) of the same party, Kenyatta of KANU (http://www.uhurukenyatta.co.ke), who later to pulled out of the presidential race in favor of Kibaki, all followed suit.

As demonstrated in table 3.1 below, out of the eight candidates who had declared their candidature for the 2007 presidential elections by June 2007,

Table 3.1 Presidential candidates' Web sites

Candidate	*Party*	*Web site/Web Address*	*Interactive facility*	*Feedback facility*	*E-mail address*	*Physical address and phone no.*
Mwai Kibaki	PNU	www.kibaki.co.ke	No	No	No	No
Raila Odinga	ODM	www.raila2007.com	No	Yes	No	No
Kalonzo Musyoka	ODM Kenya	www.kalonzomusyokaforpresident.com	No	No	No	Yes
Julia Ojiambo	ODM Kenya	www.electjulia.com	No	Yes	No	Yes
William Ruto	ODM	www.williamruto.com	No	Yes	No	Yes
Najib Balala	ODM	www.najibbalala.net	No	No	Yes	Yes
Musalia Mudavadi	ODM	www.mudavadiforkenya.com	No	Yes	Yes	Yes
Uhuru Kenyatta	KANU	www.uhurukenyatta.co.ke	No	No	No	No

only one, the then and current president, did not have a Web site. However, Kibaki was to inaugurate his Web site (www.kibaki.co.ke) on November 22, 2007, two months after launching his new vehicle to the presidency, the coalition Party of National Unity, on September 16. Only two had e-mail addresses on their Web sites by mid September 2007. None of them had any interactive facility, and only three had feedback mechanisms.

Predictably, many—if not all—of these sites become unused, or are misused by "hijackers," within months of the vote. The Web sites developed for the earlier 2002 elections lasted only about three months after the conduct of polls, and those of Ruto, Kenyatta, and Ojiambo had in fact been deactivated by April 2008, only four months after the election. This demonstrates the existing opportunistic use of the web by political elites, supporting the argument that political parties and politicians would use available communication resources for self-gain, especially in preserving or gaining their positions in power, without drawing up follow-up plans to keep the ordinary people reengaged with the political system. This is not unique to Kenya, however, and is the source of severe criticisms leveled at competitive party politics in many countries subscribing to Western "democracy." What is more, the postelection disappearance of party Web sites reflect the true nature of African political parties, which become active at election time and return to their lethargic or autocratic traditions in between polls (Murunga and Nasong'o 2006, 2007; Nyabuga 2006).

Conclusion

This chapter has offered insights into the political developments in Kenya and the use of new media—specifically cellular telephony and PC-based Internet—by presidential candidates. It has argued that while there is significant use of these media, there is little to indicate that they hold any significant advantage over traditional media forms. For this reason, they cannot be used in isolation from the "older" media. Put differently, new media are often complementary, and can only be successful when used alongside the more common forms of media, such as print and broadcast, as well as—especially in Africa—trusted face-to-face interaction. Whilst they may offer advantages in terms of accessing the electorate directly, and can circumvent information gatekeepers and political intermediaries, they may offer only limited influence on the political process. Of particular significance in this regard is their place in political communication, helping to spread the message, and sometimes promoting a two-way communication process. In short, new media can play a complimentary role, aiding communication, lobbying, and campaigning. However, such important aspects as the senders and the messages themselves, as well as people's perceptions of the candidates, play an important role in the elections and its attendant outcomes.

Thus, whilst cellular telephony and the Internet offered Kenya's 2002 and 2007 presidential candidates opportunities to stimulate participation and support, they are not yet mainly, and will not ever be solely, reliable as the means of accessing the electorate. Oftentimes, Kenya's politics is steeped in older conventions (campaign rallies, meet-the-people tours, print and broadcast coverage, or adverts), which remain the reliable means of accessing support and the electorate. Additionally, much more is required to be done to activate people's interest in politics, particularly as citizens seem to be suffering from political fatigue (cf. Nye, Zelikow and King 1997; Fullinwider 1999; Norris 2000; Webb, Farrell and Holliday 2002; Lofgren and Smith 2003; Zajonc 2004; Murunga and Nasong'o 2006, 2007). In other words, new media cannot on their own engineer political interest and activity without a significant shift in political structures, behavior, and culture to make politics more interesting, and inclusive.

Granted, irrespective of some of the above seemingly ambivalent arguments regarding the status of new media in Kenya, that there was significant use of new media in Kenya by various political actors in their attempts to reach out to the especially younger electorate, and also as channels of political communication. This conclusion is premised upon the fact that new media irrefutably offered political parties and the presidential candidates advantages over some commercial media that provide a skewed coverage. Moreover, given the fact that new media are not as vigorously controlled by the state as are "older" media, and given the capacity of the senders of the messages to formulate and manage content to communicate directly with users or consumers, new media were, and will always be, considered attractive to political elites seeking public office. Although it may not as yet be clear whether the new media influenced the outcome of the presidential elections, they offered the means for effective political communication, which is fundamental to any elections. Indeed the situation is changing slowly as younger generation of Kenyans, who grew up consuming considerable new media content, become increasingly significant politically.

Notes

1. Kibaki was later to retain the seat in a vigorously contested, but later highly disputed, poll that resulted in widespread violence and the loss of over (conservatively) 1,500 lives and displacement of thousands. It took the intervention of the international community, led by former United Nations Secretary General Kofi Annan, to resolve the disputes and quell the violence and anarchy. The chaos only subsided after the announcement of a power-sharing deal, with Kibaki as president, and Raila Odinga, the Orange Democratic Movement party frontman who was considered to have won the poll, as prime minister—a position created to facilitate the settlement.

2. See for instance, Mango and Mutua 2007. Also at *http://allafrica.com/stories/200708211233.html* (accessed July 25, 2008); Kimani, M. 2007. "Cash Becomes King in Kenya's Election Politics." *Business Daily.* October 12, 2007. http://www.bdafrica.com/index.php?option=com_content&task=view&id=3614&Itemid=5822 (accessed July 25, 2008).
3. The Security Act gave the president the power to detain without trial anybody who "threatened" state security, and numerous recalcitrant politicians, including Oginga Odinga, Raila Odinga, Kenneth Matiba, and Charles Rubia, were detained without trial. See, for example, Adar and Munyae (2001).
4. See ITU African Mobile Indicators, http://www.itu.int/osg/spu/presentations/2005/kelly-mobile-africa-5-july-05.pdf (accessed November 25, 2006).
5. See Kenya Television Network News, 9.00 PM, August 20, 2008, citing Communications Commission of Kenya sources.
6. See, for example, Ondari, J. 2007. "Firms to Bring Rural Areas Online." *The Nation*, August 1, 2007. http://www.nationmedia.com/dailynation/nmgcontententry.asp?category_id=3&newsid=103510 (accessed August 2, 2007).
7. Phombeah, G. 2001. Kenya Parties Go Online. http://news.bbc.co.uk/1/hi/world/africa/1432382.stm (accessed July 25, 2008).
8. Ibid.
9. Personal interview with Ramadhan Juma, the National Executive Officer of the Democratic Party of Kenya. Nairobi, March 22, 2005.
10. Ibid.
11. Personal interview with Moses Owuor, a manager with IFES. Nairobi, February 15, 2005.
12. Ibid.

References

Adar, K. and I. Munyae. 2001. "Human Rights Abuse in Kenya under Daniel arap Moi 1978–2001." *African Studies Quarterly* 5, 1. http://web.africa.ufl.edu/asq/v5/v5i1a1.htm (accessed February 10, 2008).

Barkan, J. D. 2004. "Kenya after Moi." *Foreign Affairs* 83: 87–100.

Barkan, J. D. 2007. "Kenya: More Than Half Full!" *Center for Strategic and International Studies* (CSIS), March 26. http://forums.csis.org/africa/?p=30#more-30 (accessed June 21, 2007).

Buckler, S. and D. Dolowitz. 2005. *Politics on the Internet.* London: Routledge.

Chadwick, A. 2006. *Internet Politics: States, Citizens, and New Communication Technologies.* New York: Oxford University Press.

Commonwealth Observer Group.1993. *The Presidential, Parliamentary and Civic Elections in Kenya, 29 December 1992.* London: Commonwealth Secretariat.

Cooke, J. 2007. "Kenya in 2007: Glass Half Empty or Full?" *Center for Strategic and International Studies (CSIS)*, March 16. http://forums.csis.org/africa/?p=29 (accessed June 21, 2007).

Davis, R. and D. M. Owen. 1998. *New Media and American Politics.* New York: Oxford University Press.

Fullinwider, R. 1999. *Civil Society, Democracy, and Civic Renewal.* Lanham, MD: Rowman and Littlefield.

Gibson, R., P. Nixon, and S. Ward. 2003. "Introduction." In *Political Parties and the Internet: Net Gain?* ed. R. K. Gibson, P. Nixon, and S. Ward, 1–10. London: Routledge.

GoK. 2004a. *National Information and Communication Technology (ICT) Policy.* Nairobi: Ministry of Information and Communications, Government of Kenya (GoK).

GoK. 2004b. *Investment Programme for the Economic Recovery Strategy for Wealth and Employment Creation 2003–2007.* Revised ed. Nairobi: Ministry of Planning and National Development, GoK

GoK. 2006. *National Information and Communications Technology (ICT) Policy.* Nairobi: Ministry of Information and Communications, GoK.

Grossman, L. 1996. *The Electronic Republic: Reshaping Democracy in the Information Age.* New York: Penguin.

Hacker, K. L. and van Dijk, J. 2000. *Digital Democracy: Issues of Theory and Practice.* London: Sage.

Hill, K. A., and J. E. Hughes. 1998. *Cyberpolitics: Citizen Activism in the Age of the Internet.* Lanham, MD: Rowman & Littlefield.

IED. 2003. *Enhancing the Electoral Process in Kenya: A Report on the Transition General Elections 2002.* Nairobi: IED.

IED, CJPC, and NCCK. 1998. *Report on the 1997 General Elections in Kenya 29–30 December.* Nairobi: IED, Catholic Justice and Peace Commission (CJPC), and National Council of Churches of Kenya (NCCK).

Kamarck, E. 2002. "Political Campaigning on the Internet: Business as Usual?" In *Governance.com: Democracy in the Information Age*, ed. E. Kamarck and J. S. Nye, 81–103. Washington, DC: Brookings Institution Press.

Kanyinga, K. 2003. "Limitations of Political Liberalization: Parties and Electoral Politics in Kenya, 1992–2002." In *The Politics of Transition in Kenya: From KANU to NARC*, ed. W. O. Oyugi, P. Wanyande, and C. Odhiambo-Mbai, 96–127. Nairobi: Heinrich Boll Foundation.

Keren, M. 2006. *Blogosphere: The New Political Arena.* Lanham, MD: Lexington Books.

Loader, B., ed. 1998. *Cyberspace Divide: Equality, Agency, and Policy in the Information Society.* London: Routledge.

Lofgren, K. and C. Smith. 2003. "Political Parties and Democracy in the Information Age." In *Political Parties and the Internet: Net Gain?* ed. R. K. Gibson, P. Nixon, and S. Ward, 39–52. London: Routledge.

Mango, C. and M. Mutua. 2007. "Kibaki's Campaign Team Goes Hi-tech." *The Standard*, August 22. http://www.eastandard.net/hm_news/news.php?articleid=1143973314 (accessed August 23, 2007).

Murunga, G. and S. Nasong'o. 2006. "Bent on Self-destruction: The Kibaki Regime in Kenya." *Journal of Contemporary African Studies* 24(1): 1–28.

Murunga, G. and S. Nasong'o, eds. 2007. *Kenya: The Struggle for Democracy.* London: Zed Books.

Norris, P. 2000. *A Virtuous Circle: Political Communications in Postindustrial Societies.* Cambridge: Cambridge University Press.

Norris, P. 2001. *Digital Divide: Civic Engagement, Information Poverty, and the Internet World.* Cambridge: Cambridge University Press.

Norris, P. 2002. "Revolution, What Revolution? The Internet and U.S. Elections, 1992–2000." In *Governance.com: Democracy in the Information age*, ed. E. Kamarck and J. S. Nye, 59–80. Washington, DC: Brookings Institution Press.

Nyabuga, G. 2006. "Our Leaders Have a Phobia for Truth to Effect Change." *The Standard*, March 23. http://www.eastandard.net/print/news.php?articleid=38384 (accessed March 24, 2006).

Nye, J., P. Zelikow, and D. King, eds. 1997. *Why People Don't Trust Government.* Cambridge, MA: Harvard University Press.

Nyerere, J. 1968. *Freedom and Socialism: Uhuru na Ujamaa.* Oxford: Oxford University Press.

Odhiambo-Mbai, C. 2003. "The Rise and Fall of the Autocratic State in Kenya." In *The Politics of Transition in Kenya: From KANU to NARC*, ed. W. O Oyugi, P. Wanyande, and C. Odhiambo-Mbai, 51–95. Nairobi: Heinrich Boll Foundation.

Osaghae, E. 2003. "Rescuing the Post-colonial State of Africa: A Reconceptualization of the Role of the Civil Society." In *Philosophy from Africa*, 2nd edn., ed. P. H. Coetzee and A. P. J. Roux, 517–527. Cape Town and Oxford: Oxford University Press.

Otenyo, E. 2002. "Internet and Democratization in Kenya: The Impact of a New Information Technology on Multiparty Elections." In *Electoral Politics in Kenya*, ed. L. Chweya, 155–172. Nairobi: Claripress.

Oyugi, W. O. 2003. "The Politics of Transition in Kenya, 1992–2003: Democratic Consolidation of Deconsolidation?" In *The Politics of Transition in Kenya: From KANU to NARC*, ed. W. O. Oyugi, P. Wanyande, and C. Odhiambo-Mbai, 345–381. Nairobi: Heinrich Boll Foundation.

Postman, N. 2004. "The Information Age: A Blessing or a Curse?" *The Harvard International Journal of Press/Politics* 9(2): 3–10.

Rheingold, H. 2000. *The Virtual Community: Homesteading on the Electronic Frontier.* Revised ed. London: The MIT Press.

Rogers, E. 2003. *Diffusion of Innovation.* 4th edn. New York: The Free Press.

Rogerson, K. 2005. "A Lot of Good Questions, a Few Good Answers: A Look at Current Research on the Internet and Politics." *Political Communication* 22(2): 237–244.

Stanyer, J. 2007. *Modern Political Communications: Mediated Politics in Uncertain Times.* Cambridge: Polity.

Stromer-Galley, J. and K. Jamieson. 2001. "Transformation of Political Leadership?" In *New Media and Politics*, ed. Barrie Axford and Richard Huggins, 172–190. London: Sage.

Thomson, A. 2004. *An Introduction to African Politics.* 2nd edn. London: Routledge.

Throup, D. and C. Hornsby. 1998. *Multi-party Politics in Kenya: The Kenyatta & Moi States & the Triumph of the System in the 1992 Election.* Oxford: J. Currey.

Tsagarousianou, R., D. Tambini, and C. Bryan. 1998. *Cyberdemocracy: Technology, Cities, and Civic Networks.* London: Routledge.

Ward, S., R. Gibson, and P. Nixon. 2003. "Parties and the internet: An overview." In *Political Parties and the Internet: Net Gain?* ed. R. K. Gibson, P. Nixon, and S. Ward, 11–38. London: Routledge.

Webb, P., D. M. Farrell, and I. Holliday. 2002. *Political Parties in Advanced Industrial Democracies.* Oxford: Oxford University Press.

CHAPTER 4

Repression, Propaganda, and Digital Resistance: New Media and Democracy in Zimbabwe

Last Moyo

Introduction

Zimbabwe has been in a state of acute political and economic crisis since 2000. Although the causes of the crisis are multifaceted, the predominant view is that there is a breakdown in the rule of law and observance of human rights by the state that is faced with overwhelming pressure from civil society, opposition parties, and the news media to embrace democratic reforms (Ranger 2005; Zaffiro 2002; Phimister and Raftopoulos 2004). The state and its critics have fought their information and ideological battles not only through the old analogue or traditional media of radio, TV, and newspapers, but also the new digital media, such as the Internet. The focus of this chapter is on the role played by the Internet in democracy in Zimbabwe. It begins by tracing the development of the Internet and the emergence of various forms of digital interaction and information sharing occasioned by the new medium in Zimbabwe. Next, the Web sites of Newsnet, Kubatana, and New Zimbabwe.com are analyzed in terms of their content and form. Newsnet is the subsidiary of the national public broadcaster, the Zimbabwe Broadcasting Holdings (ZBH). Its major task is to provide news and current affairs programming to ZBH radio, TV, and online services. Kubatana is a civic organization involved in cyber activism to highlight democracy and human rights issues in the country.

NewZimbabwe.com is the online news services—created by Zimbabwean journalists exiled in Britain—that offers alternative news and some interactive space for Zimbabweans. Critique of these Web sites' content and interactive formats is done so as to draw some inferences about their contribution to the structural and substantive elements of digital democracy in Zimbabwe. Through Critical Discourse Analysis (CDA) of online news articles—sampled through typical case sampling strategy—the chapter demonstrates how news discourse is implicated in reproducing dominant ideologies (Fairclough 1995, 2001; Wodak and Meyer 2001). The public sphere theory—described by Price (1995, 24) as an indispensable "technique for evaluating speech practices and media structures to measure progress towards a democratic society"—underpins the entire critique of the form and content of the Web sites.

Foundations of Digital Democracy in Zimbabwe

Digital democracy can be defined as "a collection of attempts to practice democracy without the limits of time, space and other physical conditions, using [new] ICT...as an addition, not a replacement, for traditional 'analogue'...practices" (Hacker and van Dijk 2000, 1). Just like in most African countries, the advent of digital media, such as the Internet, in Zimbabwe has been central in the creation of digital public spaces and public spheres for strategic institutions, such as the news media, civil society, government, parliament, and political parties—providing them with new innovative ways for public communication and information production, packaging and dissemination. For example, the Movement for Democratic Change (MDC) party extensively used the Internet for its election campaign in the 2000 legislative elections and the 2002 presidential elections, during which it depended largely on its Web site and e-mails. The then MDC webmaster, Brenda Burrel, argued that "the website allowed accurate and timely coverage of the party's activities, bypassing the misrepresentation and censorship of the local state-controlled media. It provided a wealth of information to foreign news services, human rights organizations, citizens and friends abroad" (Balancing Act, January 28, 2004). These new methods of communicating and sharing information—seen as marking the nascent stages of a digital communicative democracy in Zimbabwe—are themselves the result of a "good" national information and communication infrastructure. Zimbabwe has a relatively good telecommunications infrastructure that has not only made it part of the global information society dream, but has also acted as a foundation for digital democracy where the news media, civil society, and other organizations make different online initiatives that have varying ramifications and implications on the substantive and structural aspects of democracy (Jensen 2002 and 2004; Mazango 1998). According to the Africa Information Society Initiative (AISI), the e-mail facility was started in 1994, and the Internet followed suit three years later when the Zimbabwe government commissioned Global One to erect a national and international Internet telecommunications network link to the United States and Points of Presence (POPs) in four major cities of Harare, Bulawayo, Gweru, and Mutare (AISI Web site, April 2004). This network link was further upgraded in 1998 when it was linked to TeleGlobe in Canada to further boost its capacity. Tel One—wholly owned by the government—owns the main network for national and international telecommunications, including the Internet. It has two international digital gateway exchanges and satellite earth stations in Gweru and Harare. Powertel—a part of Zimbabwe Electricity Supply Authority—the National Railways of Zimbabwe (NRZ), and Africom Limited also provide some of the network infrastructure used by Internet service providers (ISP) to relay data and information from organizations and individual subscribers.

Zimbabwe has numerous ISPs that are quite diverse, and they potentially provide the opportunities for a digital communicative democracy for people and organizations that can afford the Internet, which is fairly accessible in terms of cost and availability. The country boasts more than fifteen ISPs, although the major players comprise Com One, M-web Zimbabwe, Ecoweb-Zimbabwe, Africa Online, Samara, Data Online, and MANGO—a number of them are present in other African countries and have global connectivity arrangements (AISI, April 24, 2004).

The rise of digital media—partly seen from the exponential growth of Web sites offering interactive spaces for Zimbabweans—can also be attributed to the number of Zimbabweans living abroad, especially in Europe and the United States. Since 2000, a significant number of Zimbabweans have left the country due to political and economic reasons. In 2004, the Zimbabwean government, through its central Reserve Bank of Zimbabwe (RBZ), estimated that approximately 3.4 million Zimbabweans had left the country to settle in countries such as the UK (1.1 million legal and 800,000 illegal), United States, Australia, Canada, New Zealand, South Africa (1.2 million), and many other African countries (BBC News, November 8, 2005). A significant number of these people are professionals such as doctors, nurses, and teachers who are likely to rely on the Internet as an information resource tool. These various diaspora communities have developed a myriad of cyber public spaces and spheres where citizens can discuss politics or read and listen to the news. Examples include, inter alia, New Zimbabwe.com, Zim Online, Inkundla, SW Radio Africa, Shaya FM, Afro Sounds FM, and TalkZimbabwe.com. Although some of these online media initially began as initiatives of Zimbabweans in the diasporas, they have developed to include the participation of residents in Zimbabwe itself. Their popularity can be seen in live forum discussions in which the main agenda is usually Zimbabwean politics. Consequently, it can be argued that there is a slow but enduring process toward a virtual nation of Zimbabweans engaging in digital public debates in time and space because of the Internet (Mano and Wendy, 2008).

The growth of the Internet and its digital public spheres has also arguably been largely motivated by Zimbabwe's authoritarian and repressive political environment. Since 2000, the government has continuously enacted laws that grossly curtail basic civil and political liberties such as the freedom of expression, opinion, association, and information. Examples include the Broadcasting Services Act (2001), Access to Information and Protection of Privacy Act (AIPPA) (2002), Public Order and Security Act (POSA) (2001), and Interceptions of Electronic Communications Act (2007). These laws and some extrajudicial tactics used by the government have not only muzzled the media and constrained civil society's political activism in "real" space, but have also arguably closed the democratic space for civic networking, mobilization, and participation in national politics. Many human rights reports claim that those who try to resist the laws have been threatened, arrested, tortured, deported, or murdered.[1] Many local journalists have been arrested or tortured and foreign correspondents deported (RSF 2002, 1). In January 2001, the *Daily News* printing press was bombed, and the offices of a prospective private radio broadcaster, Voice of the People (VOP), were bombed two years later.

As a result, public spheres for free expression and fearless political debate now find expression more through the new ICTs, as evidenced by the seemingly unprecedented mushrooming of online activities by the individual citizens and organizations since 2000. The potential political impact—real or perceived—of the new ICTs has not gone unnoticed by the government. At the World Summit of Information Society (WSIS) in 2003 in Switzerland, Mugabe openly criticized the Internet as an imperialist tool causing political conflict in Zimbabwe. The government has also gone on to try and restrict the nascent forms of digital resistance among citizens through a range of strategies. For example, fourteen people were arrested in November 2003 for circulating mass antiestablishment e-mails that were part of some efforts to mobilize people to participate in demonstrations against the government because of the rising cost of living. Since 2004, there have been some consistent reports of government intentions to purchase cyber surveillance equipment (Jackson 2004; Mukundu 2004). In 2007, the government enacted a new law called the Interceptions of Electronic Communications Act to enable it to spy on the flow of information on the Internet in Zimbabwe.

The optimism for the growth of digital democracy is also based on the number of people who use the Internet in Zimbabwe. For example, in 2002 Zimbabwe had about 400,000 Internet users, and in 2003 it had the highest growth rate of Internet users in Africa, with about 400 percent rise between 2001 and 2002 (UNDP 2003, 30–34). In June 2008, the number of Internet users was estimated at above 1.3 million (Miniwatts Marketing Group, July 10, 2008). This ever rising number of Internet users is also partly reflected in the mushrooming of cybercafes almost in every corner of Zimbabwe's major cities. In 2008, Harare alone had more than twenty cyber cafes while Bulawayo had between ten and fifteen. Apart from cyber cafes, other Internet public access points in Zimbabwe include public libraries, ten universities, and several public and private colleges. Many organizations also use the Internet to promote democracy in Zimbabwe.

Public Media Web sites: Newsnet

In July 2007, the former Minister of Information, Prof. Jonathan Moyo, made an important, but belated, admission that the Zimbabwean ruling party and government "[had] since 2000 defined the [Zimbabwean] national crisis only and always in propaganda terms" (Moyo 2007). Although this information about the abuse of the public media institutions by the ruling elite was not new to many Zimbabweans, it is important that the comment came from someone who, since 2000, had worked in the government as the kingpin of its public communication strategy for nearly five years. Apart from the fabricated stories about certain opposition leaders, propaganda[2] could also be gleaned from these media's deliberate and systematic focus on the high-sounding, yet partially hollow and misleading, grand narratives of nationhood and imperialism to frame the Zimbabwe crisis. Many examples abound.

The ideology of the nation can be seen in how a story published on ZBC's Newsnet Web site on August 9, 2004 foregrounds nationhood in its headline, "Zimbabweans urged to remember fallen heroes."[3] It is important to note that the subject is not "People" but "Zimbabweans"—framing the news in a way ideologically consistent with the agenda of the story. The narrative of the story also shows that the news is framed within a nationalistic discourse based on other constructed myths of patriotism, heroism, bravery, and sacrifice prefigured through Mugabe's reference to the "fallen heroes" who "paid the supreme price and are lying in unknown caves [and] forests." While the liberation struggle, indeed, represents the collective experience of the war by Zimbabweans, it can be argued that the reconstruction of the notions of patriotism and heroism in online journalism as a social practice is a highly ideological process that may be directly or indirectly linked to the legitimation of the prevailing political order. Hence, Hartley (1988), Koch (1990), Fairclough (1995), and Allan (2004) argue that news is a great myth or ideology maker where journalists unconsciously act as the agents.

In the story, the myth of the "Unknown Soldier," represented by the "Tomb of the Unknown Soldier" (a symbol of patriotism and sacrifice to one's country), arguably enhances the ideological narrative of the news whose agenda seems to be to take citizens as loyal subjects to the discourse of nationhood, which may be embedded with certain elite class interests at the expense of the individual political rights such as freedom of expression and the right to have different political opinions from those that define the status quo. Tamir (1999, 69) argues that the ideology of the nation is always "associated with the rhetoric of blood and soil, fiery exhortations to rally behind the flag, demands for total devotion and self sacrifice and a collective feeling that sweeps away individual considerations." Thus, the ideological function of this story seems to be not only to legitimize the ideology of "Zimbabweanness," but also to set the boundaries for alternative views and ways of thinking about the Zimbabwean crisis and its resolution. At this level, it can be argued that such news stories from the public media seemingly play the role of "narrativized ideology" through the "formulaic articulation and naturalization" of the discourse of the nation so as to frame nationalism as the only answer to the Zimbabwean crisis (see Fulton et al. 2005, 19).

Fairclough (1995, 91) argues that, as narrativized ideology, news stories tend to be targeted predominantly at working-class audiences because "stories are for those who, because of their social status and education, are denied the power of exposition, while exposition is for those who have... [power]." It is imperative to note that because of his privileged social status, Mugabe is the one who primarily defines Zimbabwe's political reality, the nature of the threat to the nation, and the nature of commitments that are needed to deal with the threat, and that Newsnet attributes significance to his definition of that reality, thus showing that, as Hartley (1988, 20) argues, "one of the primary functions of the news in [the media] is continuously to signify myths through everyday detail of newsworthy events." The emphasis on Mugabe's characterization of the Zimbabwean crisis in the story as being essentially about safeguarding national sovereignty against imperialism clearly demonstrates how power regulates discourse in terms of what can be said or not said. As Barker (2001, 12) argues, discourse is "not proliferating in an endless deferral but is regulated by power which governs not only what can be said under determinate social and cultural conditions, but who can speak, when and where."

Another Newsnet story "MDC a party of puppets—Professor Jonathan Moyo" (July 1, 2004)[4] extends the imperialism myth about Britain and other Western countries' desire to recolonize Zimbabwe and "destroy all that is Zimbabwean." A closer look at this and the earlier story shows that the ideology of nationhood and imperialism seem to be predicated on the binary oppositions of "Us" and "Them," implying strong, rigid, and natural differences between the groups. The analysis of similar articles, such as "All set for Umdala Wethu Gala," "Government remains opposed to Imperialists," "Zimbabweans only can decide," and others published in the Newsnet Web site in July 2004 shows that its news construct the nationhood and imperialism myths through creating categories and dichotomies of "Us" (i.e., the ruling party) and "Them" (all those critical of the state). The prevalence of these binary oppositions in the Newsnet stories show how it recasts the national public debate about the crisis in Zimbabwe in a segregatory way to include and exclude certain groups, and malign and praise certain other groups, in ways that render citizens as either "in-groups" or "out-groups" to the whole nationhood project. For example, the Web site seems to recast national identity in binary oppositional ways that are suggestive of enmity and hatred between not only the white and black races but also the political groups in the country.

On one hand, insiders to the myth of nationhood are described as "comrades," "the people," "revolutionaries," "patriots," "peace loving Zimbabweans," while, on the other hand, outsiders are vilified for their different political opinions through epithets such as "sell-outs," "enemies of the people," "anti-revolutionaries," "enemies from within," "enemies of the state," "detractors," "puppets," and "thieves."[5] Such binary divisions stigmatize and "demonize" the alternative thinking associated with the opposition and social movements critical of the government. They are defined as inherently evil, while the ruling party represents everything that is good and progressive. These categories are represented by Newsnet as rigid, immutable and endemic; yet ironically, they are a product of its own discourse that seem to "systematically form the objects of which [it] speaks of" (Wodak and Meyer 2003, 18). Fairclough (2003, 53) argues that in news frames "when the voice of another is incorporated into a text, there are always choices about how to 'frame' it, how to contextualize it, in terms of other parts of the text." What the journalists did here was to exclude the voices or views of the so- called imperialists and puppets while amplifying those of Mugabe and Jonathan Moyo. According to McQuail (2005, 378–379), framing is primarily about selection and salience, where the writer attributes or denies value to certain things so as to create certain meaning potentials. The stories do not raise anything about the democratic initiatives of the local Zimbabweans toward legitimate regime change, but probably overplay the imperialism issue. Ultimately, such news frames render the local and national dynamics of Zimbabwean politics as insignificant and not newsworthy, while probably exaggerating the role of the British and

Western forces in what the news constructs as the internal affairs of Zimbabwe. Through the repetitive and consistent coverage of the so-called imperial element in Zimbabwean politics, a marginalization of the news that gives salience to the initiatives of the opposition parties and other social movements in the struggle for the re-democratization of the country is created in a self perpetuating manner.

Ironically, these polarities of "Us" and "Them" seem to be arbitrary and unstable because they construct a Zimbabwean political reality that consists of categories that appear to be simple and clearly defined, yet in reality they are a subject of debate and analysis. For example, given the Zimbabwean political context, the dividing line between patriotic/unpatriotic, heroes/villains, and nationalists/sellouts, is indeed problematic and not an act of common sense as sometimes implied in the news from Newsnet. National interest, as the ultimate benchmark for the classification of these groups, is, in-itself, a blurred and restrictive concept that only appears to limit free expression and stimulate reflexive debates in the Web site. To that end, Newsnet seems to consciously or unconsciously play a divisive role by distorting citizens' political roles to create heroes and villains whose ideological and political differences are irreconcilable. As Jonathan Moyo says elsewhere on the Newsnet Web site "...there are two Zimbabweans today and these are, one, a caricature of the British, the White Commonwealth, the American and the former Rhodesian and the other one is the real Zimbabweans" (Moyo, July 23, 2004). The binary oppositions in such news appear to also privilege one category at the expense of the other because the political identities of those who criticize the state are only constructed against those of the ruling party.

As a result, the Zimbabwean opposition and other democratic forces are not represented independent of the overarching and dominant identity of the ruling party, but are only seen as representing the absence or loss of those values that characterize Zanu PF party. Consequently, once the opposition groups are constructed by Newsnet as unpatriotic, inferior, unstable, evil, irrational, and a threat to Zimbabwean nationhood, then the state can use any means of repression, such as the draconian laws and violence, to deal with them. This seriously undermines the role of the "new media" as political public spheres in Zimbabwe. The representation of the opposition as traitors arguably constructs participatory politics in Zimbabwe as "treasonous" and an "act of betrayal of one's country" for those citizens that support the opposition's policies. This amounts to the intimidation of the people who have different political opinions from that of the ruling Zanu PF party while seriously undermining the culture of deliberative democracy and public debate about issues of national importance.

As a public sphere, Newsnet is not supposed to "demonize" citizens who hold different political opinions, rather it should empower them to "exercise [some] informal control of the state through the pressure of public opinion" (Curran 1999, 20). It must be the custodian of democracy by representing views of divergent social and political forces in a "fair" and objective manner, thus creating a public sphere that is "oriented towards inter-subjective understanding and consensus" (Holub 1991, 8). The systematic use of nationhood and imperialism and the exclusion of voices that have different views to that of Mugabe in the news amounts to a highly monolithic ideological and propagandistic narrative that gives preponderance to the ideas and interests of the ruling elite. Perhaps that explains why Raftopoulos (2004, 1) argues that both discourses are simply a hegemonic project of the government, which, under the current land reform fiasco and international isolation, seeks not only to win the people's hearts and minds but also to reconstruct the postcolonial state "in order to provide both the modality for and consolidate the accumulation drive of the ruling party elite."

Alternative Media Web sites: Kubatana.net and New Zimbabwe.com

Nancy Fraser's idea of "counter public spheres" or "subaltern public spheres" is an indication that, by their very nature, discourses are always made of various competing strands that reflect the variegated nature of social interests in society. Fraser actually argues that a single, all-inclusive public sphere may be undemocratic and that the proliferation of a multiplicity of competing public spheres may be a step toward a participatory democracy in structural and substantive terms (Fraser 1992, 109).

As a civic organization, Kubatana.net represents an alternative public sphere where human rights—not nationhood or imperialism—are constructed as the epicenter of the Zimbabwean crisis. Democracy and human rights underpin most of its narratives, such as news, victim accounts, witness accounts, poetry, art, and so on. Its human rights discourse shows that the organization is, indeed, playing a crucial role in the postindependence democratization efforts in Zimbabwe. For example, the story entitled "Police Brutality in Zimbabwe" reports how the state unreasonably uses violence on "peaceful demonstrators" who have "reasonable and legitimate complaints," such as "poor wages," "high inflation," and poverty.

Like Newsnet, Kubatana.net seems to construct the human rights discourse in the binary divisions of "good" versus "evil" and "perpetrators" versus "victims." To legitimate the discourse, it is interesting to note how the story is based on the myth of life as stable, orderly, democratic, and nonviolent—myths that seem to buttress a human rights ideology that is in itself inherently problematic because of its propensity to give prominence to conflict, drama, the bizarre, and the sensational in its narratives.

In the story referred to above the trade unionists are represented as peaceful, unprovocative, and orderly while the police are constructed as unreasonable, savage, and brutal. This reinforces Kubatana's human rights and democracy perspective and has a beckoning effect on the reader so that they begin to see Mugabe and his regime as evil, violent, and undemocratic. The concentration on human rights and governance issues by civic organizations such as Kubatana.net has earned them the criticism for ignoring the issue of the social instability caused by the negative impact of the economic neoliberal reforms and the colonial legacy on Zimbabwe's sociopolitical and economic fabric (Sachikonye 2003). In most of the human rights violation accounts from Kubatana.net, there tends to be no historical context to the crisis of civil and political liberties, and this seems to suggest that Kubatana.net gives no emphasis to the fact that political rights are also dependent on a properly functional economy whose collapse might have grave results on the former. This, however, cannot be seen as entirely undermining the potential of Kubatana.net in giving alternative views a voice and a platform for digital resistance.

Kubatana.net has hyperlinks with close to 360 civil society Web sites that specialize in various fields such as democracy, human rights, elections, media, citizenship, and many other sectors. Its discourse on these issues is hypertextual, comprising the rich, intertwined, endless, open-ended, and multimodal forms of text, sound, and video situated in different sites that are part of its network. Kubatana.net not only has virtual links but also has an online telephone directory of civic organizations for use by the public. The electronic directory shows that the relationship between the hundreds of organizations in the Kubatana.net Web site goes beyond the Internet to real-life interaction. As Pauwels (2005, 605) argues, "there are many complex interactions and connections between off-line and online [public spheres], and both are part of the contemporary social life of an increasing number of people and organizations." The Kubatana.net virtual community is therefore a reflection of real life interactions between various organizations that advocate for democracy in Zimbabwe. As an online community, it is arguably the greatest virtual community that provides advocacy information on democracy and human rights issues. According to one of the directors, Brenda Burrell, it is "considered one of the best practical examples of cyber-advocacy in Southern Africa" (Interview, July 20, 2004).

Norris (2001, 186) also argues that civil society uses the Internet for, among other things, "lobbying elected representatives, public officials, and policy elites" and "networking with related associations and organizations, mobilizing organizers, activists, and members using action alerts, e-mail." Kubatana.net also uses the Internet for direct engagement with government, political activism, civic mobilization, and networking, and general information dissemination through the Web site, web logs, and mailing lists. In July 2004, Kubatana's cyber advocacy covered human rights issues such as police brutality, unfair electoral laws, unjust media, and communication laws. In most of these cases, the public were urged to take action by writing protest letters and e-mails to the authorities and to relevant civil society institutions that can represent and articulate public interest at policy level. Such initiatives by Kubatana.net may in future be harnessed to develop a sound culture of digital democracy in Zimbabwe.

Kubatana.net tried to organize mass protest e-mails and letters against the state's attempt at cyber surveillance. In a sense, it could be argued that Kubatana.net represents a potentially powerful tool for democratic participation and digital resistance for Zimbabweans who have access to the Internet. It epitomizes the apex of the political public sphere's function of not only shaping opinion but also delivering it to the concerned authorities. To a greater extent, Kubatana.net shows that, as Anheir (2003, 3) contends, civil society as public spheres are indeed "the infrastructure that is needed for the spread of democracy." Its electronic activism shows how, as Norris (2001, 97) claims, the "internet can function to strengthen and enrich connections between citizens and intermediary organizations like social movements, interest groups and governments." Indeed, Kubatana.net is a good example of how the Internet can be used to create strategic alliances among civic organizations that have a common agenda and how the Internet can be used to share information amongst its members and also engage with the government. In the words of its Director, Burrell, Kubatana.net is "a one-stop" shop or "an information hub that reflect[s] the views of Zimbabwean civic organizations" (Interview, July 20, 2004). Mercer (2004, 57) argues that although there are currently big structural and funding problems for African civic organizations such as Kubatana.net, the wisdom behind encouraging them to use the Internet is that "alliances and coalitions…[may eventually]…emerge which can hold government to account more effectively." However, it is these alliances and coalitions that Hassan (2004) criticizes at an ideological level as neoliberalistic in approach and worldview. He contends that "…we are living in the midst of a crisis of civil society stemming from the effects of neo-liberal globalization and ICT revolution" (Hassan 2004, 100). He sees civil society as advancing neoliberal free market interests, arguing that "the nexus between neo-liberal globalization and ICT revolution has spread the theory and practice of the market into…the civil society" (Hassan 2004, 102).

Just like the Kubatana.net, the ideology of NewZimbabwe.com seems to be that of activism seeking to expose political and economic bad governance and the violation of human rights in Zimbabwe. The Web site states that "we…expose the false revolutions sprouting across Zimbabwe and record the atrocities of the ruling oligarchy without fear"

(NewZimbabwe.com, April 7, 2008). This Web site is arguably the most famous among not only the Zimbabwean citizens living in the diaspora but also in Zimbabwe itself. It covers issues like politics, business, arts and culture, and sports. It has created a good reputation as a Web site that provides reliable news reports and a platform where Zimbabweans in the diasporas and at home exchange views in its discussion forums about the social and political developments and the future prospects in their country. Most of the news tends to be a collection of the news published by Zimbabwean private news media such as the *Standard*, the *Independent*, and international news agencies and the global news retailers. New Zimbabwe.com also has a vibrant discussion forum where participants from different parts of the world discuss politics and other issues pertaining to Zimbabwe. In some cases, some public officials in Zimbabwe have been invited to answer participants' questions in the forums, thus demonstrating digital democracy in action.

Conclusion

It can be argued that the Internet has not only expanded the form and pushed the physical boundaries of Zimbabwe's public spheres by providing opportunities for virtual interaction in time and space for the scattered Zimbabwean population, but also through new multimodal narrative techniques brought about by multimedia convergence. Information from the Web sites of Kubatana.net and NewZimbabwe.com often exists in multimodal texts where readers can get immersed in the search for the "truth" about events and processes in Zimbabwe. The streaming of information through audiovisual means using Windows Media Player by Kubatana.net and links to You Tube by Newzimbabwe.com promote "a more powerful sense of user engagement with media texts, a more independent relation to sources of knowledge, individualized media use, and greater user choice" (Lister et al. 2003, 20). Video and audio streaming means that users of these Web sites can watch, pause, rewind, forward, or even download a program so as to watch it at their convenience. The web log and mailing lists in Kubatana.net also encourage interaction among the users of the Web site, in the fashion defined by Lowrey (2006, 478). Users upload personal stories, pictures, and videos, or they initiate interactive discussions about their political and social experiences in Zimbabwe. Kubatana.net web logs deal with a lot of issues, such as state violence, torture of activists and opposition members—in story or debate formats—from the point of view of the people concerned. As such, Kubatana.net and New Zimabwe.com Web sites seem to balance information provision with the right of the users to communicate through web logs and discussion forums. On the other hand, the Newsnet Web site is not interactive and the users are just bombarded with huge amounts of "monomodal" information in a medium that has the potential to create multimodal texts. User interaction on the Newsnet Web site is limited to merely forwarding a story to a friend. The Web site has no discussion forums and online streaming facilities, especially to cater for the diverse interests and differing abilities of its audiences, such as the blind.

The analysis of the Web sites' content shows that the use of the Internet as a new medium remains largely in the hands of the powerful in society, where the state and donor-funded elite institutions consciously or unconsciously use it to promote discourses that revolve around their own interests. As Preston (2001, 209) contends, history shows that "... entrenched economic and political interests... readily, adapt and harness the internet in line with elite interests." While the Zimbabwean government has every right to defend its sovereignty online, Newsnet reflects an entrenched culture of authoritarianism that is presented in a popular idiom of nationalism. As such, the Internet can neither break new vistas for democracy nor create participatory politics outside the overarching interests of the elite. Hence the observation by Slevin (2000, 155), that "...any meaningful analysis of the impact of the internet on society...must also be fundamentally cultural," holds true.

The potential of the Internet is also limited by the new media laws in Zimbabwe. The Postal and Telecommunications Regulatory Authority of Zimbabwe (POTRAZ) seems to be unable to operate autonomously because of government control. The Transport and Communications Minister solely appoints the Board of Directors, accused by Zimbabwe ISPs of being government loyalists (Holland, interview with the author, September 11, 2004). According to Jim Holland, POTRAZ does not seem to be protecting the interests of the telecommunications sector, but instead acts as a gatekeeper for state interests in the sector. It has also been accused of stifling growth and competition in the new ICTs. For example, telecommunications in Zimbabwe remains largely a state monopoly, where the state-owned Tel One not only effectively monopolizes the provision of leased lines for Internet access but also owns the main national and international network that most of the ISPs use for their data transmission.

Apart from regulatory problems, digital democracy is also inhibited by factors such as affordability, availability and accessibility of the hardware, software, and a robust telecommunications infrastructure. For instance, while the e-mail and chat forums are undoubtedly a central part of Zimbabwe's nascent digital democracy, the Internet is still, relatively, for affluent Zimbabweans and those in the diaspora. Apart from these, the typical cybercafé user in Zimbabwe does not seem to have a sustained use of the Internet to engage in forum discussions because of the stated inhibitions.

Notes

1. Various reports by Human Rights Watch, Human rights Trust, Amnesty International, Legal Resources Foundation, and Africa Union indicate that opposition supporters were tortured or murdered by the army, police, or people sympathetic to the state.
2. Between 2000 and 2003, the state media published a number of stories that claimed that the opposition party was planning to bomb the tallest buildings in the country and had also sent anthrax-laced letters to the state media, such as the *Chronicle*. These claims were later retracted and an apology was published to the MDC.
3. See http://www.newsnet.co.zw/index.php?nID=2042 (accessed September 16, 2008).
4. See http://www.newsnet.co.zw/index.php?nID=1041 (accessed September 16, 2008).
5. These binary divisions were not limited to only Newsnet, as they could also be found in other public media such as the *Herald*, *Sunday Mail*, *Chronicle*, and Power FM.

References

Allan, S. 2004. *News culture*. London: Open University.

Anheier, H., Glasius, M., and Kaldor, M. eds. 2001. *Global Civil Society*. Oxford: Oxford University Press.

Barker, C. 2000. *Cultural Studies: Theory and Practice*. London: Sage.

Fairclough, N. 1995. *Media Discourse*. London: Arnold.

———. 2001. *Analyzing Discourse*. London: Routledge.

Fraser, N. 1992. "Rethinking the Public Sphere." In *Habermas and the Public Sphere*, ed. C. Calhoun. London: MIT.

Fulton, H., Huisman, R., Murphet, J., and Dunn, A. eds. 2006. *Narrative and Media*. New York: Cambridge University Press.

Jackson, P. 2004. "E-mail Controls Loom in Zimbabwe." *BBC News*, May 31, 2004.

Jensen, M. 2002. The African Internet: A Status Report. African Internet Connectivity, September. http://www3.sn.apc.org/africa.

________. 2004. "The Rise of The Telecenters and Cyber Cafés in Africa." *Acacia Initiative* 9 September 2004. http://www.acacia.org.za/jensen_articles.html.

Koch, T. 1990. *The News As Myth: Fact and Context in Journalism*. New York: Greenwood.

Lister, M., Dovey, J., Giddings, S., Grant, I., and Kelly, K. eds. 2003. *New Media: A Critical Introduction*. London: Routledge.

Lowrey, W. 2006. "Mapping the Journalism-Blogging Relationship." *Journalism* 7 (4): 477–500.

Mano, W. and Willems, W. (2008), 'Emerging communities, emerging media: the case of a Zimbabwean nurse in the British Big Brother show', *Critical Arts*, 22 (1): 101–128.

Mazango, E. 1998. Telecommunications Sector Reform: Liberalization and Universal Service Policy in Zimbabwe. Diss., University of Oslo. [Unpublished]

McQuail, D. 2005. *Mass Communication Theory*, 5th ed. London: Sage.

Mercer, C. 2004. "Engineering Civil Society: ICT in Tanzania." *Review of African Political Economy* 99: 49–64.

Moyo, J. 2007. "Price War: Zanu PF Comes to the End of the Road." *Zimbabwe Independent Newspaper*, July 6, 2007. http://www.theindependent.co.zw.

Pauwels, L. 2005. "Websites as Visual and Multimodal Cultural Expressions: Opportunities and Issues of Online Hybrid Media Research." *Media, Culture and Society* 27 (4): 604–613.

Phimister, I. and Raftopoulos, B. 2004. "Mugabe, Mbeki and the Politics of Anti-Imperialism." *Review of African Political Economy* 101: 385–400.

Preston, P. 2001. *Reshaping Communications: Technology, Information, and Social Change*, London: Sage.

Price, M. 1995. *Television, the Public Sphere and National Identity*, Clarendon Press.

Ranger, T. 2005. "The Rise of Patriotic Journalism in Zimbabwe and its Possible Implications." *Westminster Papers in Communication and Culture* 2 (1): 8–17.

Sachikonye, L. M. 2003. "From Growth with Equity to Fast Track Reform: Zimbabwe's Land Question." *Review of African Political Economy* 96: 227–240.

Wodak, R. and M. Meyer. 2004. *Methods of Critical Discourse Analysis*, London: Sage.

Zaffiro, J. 2002. *Media and Democracy in Africa: Zimbabwe, 1931–2002*. Colorado Springs: International Academic Publishers Ltd.

Selected Web sites

AISI, http://www.uneca.org/aisi.

Balancing Act News Update, http://www.balancingact-africa.com.

Econet Wireless, Zimbabwe, http://www.econet.co.zw.

Economic Commission for Africa, http://www.uneca.org.

Miniwatts Marketing Group, Internet World Statistics, http://www.internetworldstats.com.

International Telecommunications Union http://www.itu.int/home/index.html.

Telecel Zimbabwe http://www.telecel.co.zw.

Tel One http://www.telone.co.zw.

CHAPTER 5

Democratic Process, Civic Consciousness, and the Internet in Francophone Africa

Marie-Soleil Frère and Alain Kiyindou

In the early- to mid-1990s, Francophone African countries went through various democratic changes, mainly based on the liberalization of political and media landscapes, the adoption of new constitutions and institutions, and pluralist elections.[1] Even though the opening up of the political processes has not been successful everywhere and the transitions have sometimes been disrupted by political leaders in an attempt to maintain power, a major step has been made in the field of citizen participation. Citizens can now speak more freely, criticize the government, question the authorities' decisions and behavior, publish and broadcast, and participate openly in civil society organizations. Freedom of expression has allowed contradictory views to be voiced in the public sphere.

Pluralist elections have been organized in each country, with varying degrees of transparency (Levine 2004). Some were used only as a democratic "façade," but at least they took place. Furthermore, a decentralization process, promoted by foreign donors, aims at bringing the administration closer to the population in order to give people more say in the decision-making process.

The role of the information media, particularly the private press and broadcasters, in this process of political opening and the advent of a new democratic citizen-identity in Francophone Africa, has been widely underlined (Tudesq 1998; Frère 2000). However, the part played by the Internet, introduced in most of Francophone Africa at the end of the

1990s, has drawn less attention (Cheneau-Loquay 2000; Ba 2003). ICTs have spread in an environment that was already transformed by liberalization of politics, the news media, and civil society, but they have brought additional changes in the fields of economy, culture, research, education, and politics.

The first section of this chapter shows how the use of computer-based Internet[2] has enabled changes in the relationship between citizens and public authorities in a number of ways: allowing more critical public debates; providing more access to official information; connecting local NGOs with civil society organizations abroad; enabling citizens to participate in worldwide virtual communities. The second half of this chapter[3] takes a more pessimistic view, emphasizing the negative consequences of the Internet in the African public sphere: cultural domination and dependency toward Western funds and strategies that do not empower citizens. Therefore, the question is whether the Internet is a new "myth" on which to ground "development efforts" or whether it actually contributes to building a more democratic and participative society in Francophone Africa.

Based exclusively on references in French, this chapter attempts to underline the main debates that were (and still are) going on within the Francophone community of scientific research about the issue of the Internet in Africa. Indeed, these debates may offer an original contribution in a research field that is largely dominated by the Anglophone approach.

ICTs Strengthening Civic Consciousness in Francophone Africa

The Internet is sometimes referred to as the "fifth estate" (Crouzet 2007), because the web gives its users a greater capacity to judge elected representatives, and also the quality of public services. Therefore, it increases the capacity of citizens to take action if they are not satisfied.

In Francophone Africa, the contribution of the Internet to strengthening civic behavior and civil society organizations is undeniable, but it is still limited. Ten years after the emergence of the Internet, the local capacity of the network is still very weak, although the situation varies from one country to another. The 2008 data from the International Telecommunication Union (ITU) puts Senegal way ahead with 6.62 Internet users for every 100 inhabitants, followed by Gabon (5.76/100) and Togo (5.07/100). Two of the main economic poles, Côte d'Ivoire and Cameroon, reach, respectively, only 1.63 and 2.23 users for 100 inhabitants. Except for Benin (1.66/100), Congo (1.70/100), and Rwanda (1.08/100), all the other countries (Burkina Faso, the Central

African Republic, Chad, Mali, Niger, Mauritania) do not even have one user for 100 inhabitants (ITU 2008).

The obstacles to a more widespread use of the Internet remain important in all fields: technical (no access to electricity or technology), economic (very high cost of network extension and maintenance), and social (high rates of illiteracy) (Bernard 2003, 127). In countries where the average income is little over $2.3 a day, a monthly package of 20 hours of Internet costs more than $50, about 70 percent of the average capita income (ITU 2008, 10). Despite these difficulties, the Internet has had some impact on civic consciousness.

Web sites Set Up by the Administration to Better Inform Citizens

The first possible contribution of ICTs to better governance and improvement of civic awareness relates to the development of the public information available in a context where administrative and official data have long been kept secret. Currently, many public services in Francophone Africa have a Web site that gives access to information on legislation, lists of staff, and daily events in the country. In Burkina Faso, the Web site of the Prime Minister (www.primature.gov.bf) presents many fundamental laws, the composition of the government, and the archives of all reports of the council of ministers. It also provides links to all the Web sites of the other government departments and public institutions. The Web site of the Senegalese government (www.gouv.sn) also includes a link called "Your Administrative Procedures," giving details about all the procedures needed for obtaining official documents and application forms related to medical care, schooling, judiciary, or registry procedures. The Web site of the Presidency of Mali (www.koulouba.pr.ml) publishes brief guides to all public institutions. In Congo, the official national Web site (www.congo-siteportail.info) provides daily news about the institutions and the activities of the government.

Such institutional Web sites can also play an important role during elections, posting online candidate lists or poll results, progressively published for each polling station, thus minimizing the possibility of fraud. In Burkina Faso, the Web site of the Independent Electoral Commission (www.ceni.bf) gave access to the detailed results from each province during the parliamentary election of 2007 and the 2005 presidential election.

Nevertheless, there are very few updated government Web sites presenting complete information and which are accessible at all times. Even though registry or administrative procedures may be described, very few Web sites allow the downloading of official forms. The Web site of the Presidency of Togo (http://presidence-togo.com) explains the role of a national ombudsman, but online submission of complaints is not possible.

In a very few countries, there is online access provided to the *Journal Officiel*, the printed version of all the legislation adopted by Parliament, which is crucial to monitor Parliament's activity.

Online Press: Raising Civic Consciousness and Democratic Debate

Another important contribution of the Internet in consolidating the civic consciousness comes from its capacity to disseminate the content of local media far beyond geographical limitation, in countries with poor roads and rail infrastructures and with mostly local FM radio stations. The press in Francophone Africa are increasingly going online, some of them, such as Dakar's *Le Soleil* (www.lesoleil.sn), are given as good examples on the continent (Pare 2007, 122).[4]

While online media play a part in the circulation of information, few of them offer the opportunity for online forum or debates. Indeed, the lack of web-trained journalists or permanent webmasters in most newsrooms limits interactivity in media Web sites.

The reasons for limiting the investment in online multimedia news are both financial and related to a lack of Internet strategy. Most newspapers just post online the same content as that which is published in the paper edition, with no changes or technical sophistication. For most media companies, the cost of getting an Internet connection or paying to have a Web site hosted on a server somewhere is too expensive. The former editor in chief of *Le Soleil* online admitted that updating and managing the Web site could be undertaken because of the personal commitment of a few journalists rather than owing to any planned strategy (Sissouma 2003, 96).

The aim of newspapers' and radio stations' Web sites is not to provide online content that is complementary or additional to the paper or broadcast information: the purpose is to make their papers and programs accessible to the diasporas abroad and to local audiences living in other cities or regions that are out of reach of paper or airwaves. Therefore, there is no need to create an online media that would provide content different from the original print or broadcast media.

Furthermore, Francophone African media are poorly indexed on the Internet and cannot be found easily, unless the Internet user already knows exactly what (s)he is looking for. Even when the newspapers or radio stations have some archives available online, they do not have a search tool.

The basic and static nature of these media Web sites makes it very difficult for them to generate profits. As the online version mainly reaches remote diasporas keen to follow what is going on in their countries of origin, the targeted readership at such distances is of very little interest for local advertisers. In Senegal, *Le Soleil* could get advertising from a local real estate company offering Senegalese diasporas the chance to invest in buying land and property in their home country. Some airlines or express money transfer companies are also interested in reaching diasporas. But the major advertisers in the local press (mobile phone companies and breweries) do not see why they should invest in reaching an audience that is not living or consuming locally. Moreover, many of those press Web sites are accessed by the Internet user through general portals that present information about a country and provide links to all local online newspapers (www.abdijan.net; www.lefaso.net). The content of the local papers is often used and reproduced without any formal agreement and it is the portal that gets the advertisements.

Even if they are not financially sustainable, the online media have helped revive a sense of national belonging among the diasporas, enabling them to join local debates and make their voices heard. Diasporas have often developed perspectives and points of view that are different and constructive, as these are enriched by their experience abroad.

Locally, only a few happy Internet users access this material. Yet the population may turn to it en masse when a particular edition happens to be sold out or when local papers have been censored or suspended and the forbidden content is hosted on the Web sites of international organizations defending press freedom. In Mali, in June 2007, the Malian audience had to log onto the Paris-based Reporters without Borders' Web site[5] to find an article, by Seydina Oumar Diarra, published in *Info Matin* (which could not be found in Bamako), which provoked a serious conflict with the authorities. The journalist and two colleagues (Sambi Toure and Hameye Cissé) ended up in jail for covering police action against a school teacher arrested for asking his pupils to analyze a short story describing a president chasing women.

International Cyber-civic-consciousness

To anybody surfing the web, the Internet provides access to a supranational digital sphere that allows the constitution of virtual communities and border-crossing networks. For instance, French President Sarkozy's speech in Dakar on July 26, 2007,[6] thought by many to have had a paternalistic and racist tone, was posted on many Web sites and sent to mailing lists and mail accounts of all the intellectual elite of the continent, which provoked a general outcry. Some of the reactions were printed in widely circulated French papers and also posted on the Web sites of Senegalese, Cameroonian, Burkinabè, Congolese, and Malian newspapers.

The former state monopoly on the circulation of news coming from abroad[7] is no longer possible. The Internet has now replaced the clandestine leaflets and photocopies that used to be circulated in the 1970s under one-party rule and has undermined leaders' plans to monopolize information to shape people's consciousness. In the Republic of Congo, where most of the local media are under government control, overseas information Web sites are the ones that provide open and free political debate.[8]

Nowadays, African governments have hardly any grip on the choices of the Internet user-consumer, who can freely choose the information that is interesting or useful and decide to join a particular "virtual community," a community to which the local citizen may not be "belonging" but will become "affiliated" (Ba 2003, 8). The only power that remains with the State is providing access to the network (Ba 2003, 245).

These "virtual affiliation communities" are centered mainly on the individual user and built through the use of e-mails, chat, "meeting" Web sites, Facebook-style sites, and personal weblogs. In Burkina Faso, a young female journalist Ramata Sore has developed her own blog, titled "Politically Committed Journalist," on which she openly challenges local authorities, denouncing the mouthpiece of politicians and calling for solidarity among all women (http://ramses1.blog4ever.com/blog). Every article she posts online gets at least a dozen reactions from all over the world.[9]

The participation in this "worldwide civic consciousness" by individuals is nevertheless linked to a double issue: access and education. On the technical side, many foreign supported initiatives have contributed to the implementation of ICTs on the national level in Francophone Africa (Heuzé 2007, 44–51). Many development NGOs have supported the establishment of public collective access centers such as Internet cafés, Multimedia Community Centers, and call-centers (Lohento 2006, 2). Among the factors cited earlier, computer and French language literacy[10] is significant. In community multimedia centers, a "mediator" between the user and the computer is often necessary. This role is often taken up by young school learners, or "modern scribes" (Cheneau-Loquay 2000, 277). Though they can use a computer, they sometimes lack critical judgment. Lack of literacy makes one vulnerable to online swindlers. In Francophone Africa, there are countless cases of fraud committed via the Internet, through common scam e-mails.

A Well-Connected Civil Society

Beyond individual uses, the Internet has given more capacity to local civil society associations and organizations to exchange concerns and ideas with counterparts abroad. Pressure groups, including cultural minorities, have become more visible and, therefore, generally are more able to make themselves heard, thanks to the web.

For instance, local associations connect easily with wide networks defending human rights (such as the International Federation for Human

Rights—FIDH, www.fidh.org), freedom of expression (International Freedom of Expression Exchange—IFEX, www.ifex.org), or workers' rights (International Trade Union Confederation—ITUC, www.ituc-csi.org). The Internet has increased the ability of these networks to mobilize their members, to set up joint campaigns, and thereby strengthening the weaker members of the network. For instance, in 2006, Christian Mounzéo, general secretary of a Congolese association defending human rights (RPDH), was arrested twice as he was campaigning against corruption in the Congo Brazzaville as an active member of the international "Publish what you pay" network (www.publishwhatyoupay.org). His arrest was instantly denounced worldwide, and there was a fierce online debate between his Defense Council and the General Prosecutor of the Republic of Congo about his case.

The dynamics of the World Social Forum (WSF) mainly exist thanks to quick, worldwide, and inexpensive circulation of information among small organizations. Organizing the WSF in Bamako (January 19–23, 2006) necessitated the extensive use of the Internet on a continent where there are relatively few telephone lines, no good postal mail services, and where the costs of cell phone calls remain high.[11] Thanks to the Internet, many Francophone African countries now have their own local organizations or affiliates fighting causes such as the abolition of Third World debt (CADTM) or for the taxation of financial transactions (ATTAC)[12]—even though they still have limited impact.

While only a minority of the population is involved in these new global networks, their contribution to democracy cannot be underestimated. In Francophone Africa, these mediating structures are often either dismissed or nonexistent. Trade unions are weak (in countries where most of the economy is informal), political parties have lost credibility, and local NGOs are often seen as the personal business of individuals rather than promoters of community interests. Reinforcing those mediating structures, through more visibility and links to international networks, is therefore of crucial importance for the consolidation of more participatory and representative regimes. The Internet is a useful tool to facilitate such global solidarity.

The Internet, Development and Democracy in Francophone Africa: From Myth to Reality

Although the Internet is considered a tool for social mobilization, it could also standardize knowledge and culture. Presented as a tool for empowerment of people and cultures, the Internet, on one hand, aids individual emancipation and collective solidarity, on the other hand, it conveys cultural domination of the West on African ways of living and thinking. One can question, therefore, whether the Internet truly helps to build a new democratic civic consciousness in Francophone Africa if it is a tool for foreign cultural domination. It may be a tool for "empowerment," but its capacity to "disempower" should also be interrogated. In the second half of this chapter, that undesirable tendency of the Internet is presented.

The Internet: A Tool for "Acculturation"?

It is clear that "the information provided by the Internet is largely dominated by Western content" and that African knowledge is widely underrepresented "on a qualitative as well as quantitative basis" (Kiyindou 2006, 6). In the realms of culture, for instance, the web provides very little content related to Francophone Africa, which is limited to a few Web sites belonging to prominent local cultural institutions (The Ouagadougou Film Festival—FESPACO, www.fespaco.bf) or of Northern journalists interested in Africa (www.afrik.com, www.Africultures.com). Some active diasporas in Western countries have also created Web sites aiming at promoting or livening up African cultures and traditions. Research on the arts from the Mandingues (peoples spread over Mali, Guinea, Senegal, and Cote d'Ivoire) leads to the Web site of a Paris-based association organizing dance and music workshops (www.les-arts-mandingues.com) (Heuzé 2007, 82). The portal of the Soninke people (living in Mali, Mauritania, Senegal, Gambia, Bissao-Guinea, and Burkina Faso), posting online radio programs in French and Soninke and giving basic language training, is also hosted in France (www.soninkara.com) and most contributions are from the various diasporas.

Some artists have created their own Web sites to show their pieces and even to sell them online. The first Internet café in West Africa (Metissacana) was created in Dakar by the Senegalese dress designer Oumou Sy who has always claimed that the best promotion of her models was done online (www.metissacana.sn/sites/oumousy) (Cheneau-Loquay 2004a, 359). Many Web sites are also created by tourists or continent lovers featuring their own travel pictures and stories. Just as there are very few (and very poorly state-funded) museums in Francophone African countries aiming at presenting and preserving the local history and cultural past and present, no attempts have been made by public institutions to use the Internet as a window for local cultures.

Accusations of "acculturation" and "alienation" (Kiyindou 2006, 7) of African cultures through the Internet also hint at the predominance of French content on the web when it comes to former French and Belgian colonies. In order to make the information accessible, most Web sites presenting African heritage are using an "international" language, thereby excluding the many local languages. The domination of the former colonial language goes back to the "assimilation" policy under colonial rule. In every former French colony (except for Mauritania), French has remained the official language—yet "every language is carrying a culture" (Essono 2004, 316). Preserving local languages is a big issue, especially in a global world where the Internet has a tendency to homogenize expressions. For this reason, some Congolese Web sites (www.bilenge.com; www.mwinda.org, www.ne-kongo.net), even though using French, have chosen names that refer to local identity. They also use drawings and pictures (statues, flags, masks, traditional sketches, and so on) taken from local tradition.

For some authors, the question of "acculturation" goes beyond the cultural content on the Internet. Research has pointed out that the uses of the Internet, mainly formatted by and for Western users, are not appropriate to some aspects of African social customs and behaviors. Indeed, "methodologies and services proposed (by the Internet) have been developed for other cultures" (Renaud 2000, 99). Nevertheless, it is worth noting that the changes brought by the Internet are only the extension of previous modifications. Since colonial time, knowledge based on orality and the collective sphere of information has been displaced by the ability to read and write. The Internet is therefore bringing a new extension to those changes that have already disrupted the traditional hierarchies of African societies. Knowledge is changing from being the privilege of elders to being the common domain of young people, as they now have access to a direct and wide network of information and knowledge from all over the world. "If knowing means being in power, gerontocracy is tending to become 'webcacy'" (Mbengue, cited by Heuzé 2007, 97). Elders are increasingly getting dependent on the technical and linguistic abilities of the younger generation, which is obvious in community multimedia centers where they cannot do without those "mediators." Such a situation is not necessarily perceived negatively. The Internet has thus also created new relationships between tradition and modernity (Renaud 2004, 1) and also created a new motivation for learning to read and write.

Education and Training: The Road toward the West?

Education and training are of crucial importance for building civic consciousness and development, as they develop human skills able to identify important issues and find solutions appropriate to the local context. Research and university education in Francophone Africa has long suffered from serious isolation from international scientific community and from a desperate lack of documentation. Those obstacles prevented the development of new local knowledge that could be internationally recognized and valued (Seck 2000, 386).

Nowadays, the Internet is available in all Francophone universities in Africa for teachers, researchers, and students, thanks mainly to foreign aid projects. The RESAFAD[13] program, implemented by the French Ministry of foreign affairs, created centers in all Francophone national universities in Africa, giving students and teachers a free access to multimedia resources and e-learning opportunities. Collaborating with the Ministry of Higher Education in Senegal, RESAFAD has helped create a Web site (www.examen.sn) that provides pupils and students in the final year of elementary and secondary schools with the documentation they need to prepare for their final exams. Such initiatives help to compensate for the weaknesses of the local education system, as well as to motivate teachers. However, it is not a solution to those problems.

The African Virtual University (AVU, www.avu.org), launched by the World Bank, was initiated in 1997 to help African universities "to educate a new generation... " (Bonjawo 2002, 70). A decade later, the contents for e-teaching are still mostly provided by Western universities, which is not always adapted to the local context. Renaud (2005) even questions the independence of the courses from the World Bank or private companies funding the project (such as Microsoft).[14] He argues that these stakeholders could be supporting their own political and economic interests. Loiret (2005) adds that the AVU project is "a pretext to settle a sphere of domination (...) and of influence that has not been set up to integrate itself in the social African context but to transform that context despite its will." Therefore, such a program fuels the debate around the "Westernization" of the African continent.

Finally, it is also obvious that Francophone Africa is facing a growing lack of specialists in many fields. Brain drain has increased because of the lack of local means available for research and teaching. The Internet has recently reinforced the wave of departures, by largely circulating calls for candidates to teach in Western universities. The consequence of the exodus of the best educated people amounts to an estimated $ 4 billion that has to be spent every year to bring to the continent specialists from abroad to cover the gap (Tebeje 2005, 1). Given the weaknesses of the local financial resources, it is again the foreign aid departments of Western countries and international donors that provide the funding (and the teachers) to compensate for the absence.

The Internet: New Sesame for Development?

For about fifteen years, the Internet has been presented by the World Bank, the United Nations agencies, and most Western foreign aid programs as a new opportunity to push Africa on the way to wealth and development (Tshimbulu 2001, 157. For UNDP (cited in Laborde 2006, 1), "the countries that manage to make the best use of the ICTs can expect a largely increased economic growth, better social welfare and more democratic forms of government." Current policies putting forward the ICTs as a sesame to development in Africa raise two remarks among the French ICT specialists. On one hand, they reconnect with the technological determinism that has driven all the transfer of technologies to Africa since colonial times (from rural radio to education TV and computers) as a legacy of the theory of modernization and the diffusion of innovations (Mignot-Lefèvre 2006, 3). On the other hand, they are obviously leaning on a political will of integrating the African continent in the global economy, as if the crucial issue for the future of the continent was economic growth and not political participation and democratization.

Beyond such promises as e-commerce, it is difficult to prove and to evaluate the real impact of the Internet on the economic "development" in Francophone Africa, as even the ITU admitted in its 2006 report. Reports about the use of public Internet in Cameroon show that Internet use remains focused on personal e-mails and documentary research for students and teachers, strengthening communication abilities and individual social performances much more than economic capacities (Koudjou-Talla 2007, 33). Even though Internet use has grown tremendously after 1995, there is no evidence that the sector's revenue in the African continent has put the population in a better position, making people more able to participate in public debate.

Citizens without a State

Convinced that ICTs pave the way for Africa's development, the World Bank, the United Nations, and the ITU advocate liberalization, privatization, and deregulation (ITU 2008, 35; Elkyn 2001, 22). Following these suggestions, most Francophone African countries have opened up their telecommunication markets to competition. Domination by foreign investors and reduction of state role raises many questions among Francophone academics. Fullsack (2002, 14) argues ITU "has contributed to progressively erase the public services." To Mezouaghi (2004, 39), suppressing public services and policies and opening the door to foreign multinational companies only weaken the local economy. In a powerless state, with an extrovert economy mainly in the hands of foreign investors, the local citizens will have less and less grip on the choices made locally in terms of infrastructures and development. If the Internet is an opportunity to give people more voice, one may question to whom the civil society will be able to turn its voice if the local authorities are no longer the policy makers.

There is thus, among French researchers, a constant debate about the boundaries between myth and reality: the myth of the contribution of the Internet to economic development, and the reality of the economic indicators on the continent; the myth of the uses of ICTs promoted by the donors, and the reality of what the African population is actually making with this new tool; the myth of empowerment, and the reality of the lack of power.

Conclusion: What is at Stake for a New Public Civic Sphere in Francophone Africa?

The Internet may definitely be used to broaden civic consciousness, consolidate democracy, and promote local cultural values in Francophone Africa, but its expansion still faces many challenges. Most research conducted in Francophone Africa concludes that these challenges cannot be taken up without the commitment of the local authorities to invest in improving Internet accessibility. "The internet can help develop democratic culture (...) if and only if the economic and political powers in those countries decide and finance literacy programs, the expansion of electricity and main line telephone networks, etc" (Benhamou, cited in Heuzé 2007, 104). The state has a prominent role to play to provide these "public services."

The power network is concentrated in major cities and is very unstable. The main telephone lines are in bad condition and very unevenly scattered in the African region[15] and inside individual countries, making it unavoidable, in most places, to relay on costly satellite equipment. Access to broadband is still limited in a global context, even though the United Nations Conference on Trade and Development (UNCTAD) has argued, in its 2006 report, that broadband is now vital for companies and has compared it to basic survival goods such as water and electricity. While the intervention of the state is required, most research also acknowledges that it is arguable whether the state should allot the available budget for developing telecommunication infrastructures in places where the needs for water and electricity are not met.

The other important challenge rests on the commitment of local social movements, which should not lose faith in their capacity to have an impact on their environment. Recent elections in Francophone Africa, even in a country like Mali that has been shown as an example of successful transition to democracy, have revealed a major lack of interest among the electorate. There is a "democratic disillusionment" (Perrineau 2003) grounded on the conviction that democracy only improved the situation of the rich and that voting is pointless.

There is therefore a risk that the Internet will only reinforce the interests of the well-connected literate elite living in the cities, while most of the population turns their back on political participation. To Cheneau-Loquay (2002, 117), ICTs are becoming "the new discriminating factor distinguishing the poor and the rich at the international as well as at the personal levels." The Internet will contribute to creating a real democratic public sphere only if the rich, educated, and connected elite use it as much to promote the collective good and well-being of the general population (integrating the preoccupations of the poorest) as for the consolidation and promotion of their individual interests.

In 2002, Brunet et al. underlined the fact that Internet uses, in Francophone Africa, were full of paradoxes: allowing greater access to information while reinforcing exclusion and inequality; enabling a cultural opening on the world while giving access to ideologically perverse content that tends to manipulate less educated minds; requiring financial investment for its expansion while other basic needs remain unfulfilled.

Some other paradoxes have also appeared in this chapter, which have underlined the debates among Francophone academics. The Internet can be used to reinforce civic consciousness locally, but it is also a tool of globalization that leaves less and less power to local civil society to influence decision makers. It allows a growing input from the diasporas into local debates and processes, while also facilitating brain drain. It enables Africa and its people to take part in global debates and make their own viewpoints heard, but this is made possible mainly due to foreign financial support provided by international companies or institutional donors that have their own agenda, and is also in the frame of foreign-designed projects.

Between a "myth" that is grounded on real cases of empowerment of African citizens and a "reality" that is the one of low economic indicators of the continent and growing inequality, the final word belongs to the users: African users will be the ones deciding what they want to do with the Internet and for what purpose.

Notes

1. These seventeen countries, which inherited French language from French or Belgian colonial rule, are often split according to two specific areas. Francophone West Africa includes Benin, Burkina Faso, Côte d'Ivoire, Guinea, Mali, Niger, Senegal, Togo, and Mauritania. Francophone Central Africa covers Burundi, Cameroon, Central African Republic, Chad, Gabon, the Republic of Congo, the Democratic Republic of Congo, and Rwanda.
2. This chapter focuses on computer-based Internet rather than cell phone Internet, which has yet to catch on in Francophone Africa.

3. The second half of this chapter is mainly based on material that was used in a different perspective in a paper published in French by Frère M.-S. and Heuzé F. 2007.
4. In 1998, the *Soleil* Web site was mentioned among the ten most attractive press Web sites in the world by the weekly *Courrier International* (Pare: 2007).
5. The Reporters without Borders' Web site has hosted many censored or forbidden newspapers from Francophone Africa, for instance, *La Griffe* (Gabon) (www.rsf.org).
6. Available on the French Presidency Web site : www.elysee.fr.
7. Mainly through a unique State press agency that had a monopoly on receiving and dispatching the features of the foreign international press agencies.
8. See, for instance, the very politically oriented Web sites of the Congolese diaspora (www.fcd.ras.eu.org) that are close to the political opposition.
9. Ramata Sore's case is quite unusual in a country where women's voices are subdued and where very few of them work in the media.
10. In Francophone West Africa, the official literacy rate is not above 50 percent, and is even below 30 percent for Niger, Burkina Faso, Mali, and Guinea (UNDP *World Report on Human Development* 2006).
11. In Mali, fixed telephone lines have increased only by 8.5 percent from 2002 to 2007, reaching 0.69 line for 100 inhabitants. During the same period of time, Internet users have increased by 32 percent, reaching 0.81 users for 100 inhabitants. In 2007, 100 minutes of prepaid cellular phone use cost US$ 28.79, which represents 78.53 percent of GNI per capita.
12. CADTM is the Committee for the Abolition of Third World Debt (http://www.cadtm.org) and ATTAC stands for Association for the Taxation of Financial Transactions to Aid Citizens (http://www.attac.org)
13. Réseau africain de formation et d'action à distance (African network of distance training and action) is a project that equipped and trained universities and school teachers (about 3,500), mainly in Francophone countries, from 1997 to 2004.
14. Today, the AVU has got an international nonprofit organization status, but the project is still dependent on the World Bank and the African Development Bank funding.
15. The number of main fixed lines vary from 2.59 (for 100 inhabitants) in Gabon and 0.02 in the Democratic Republic of Congo. Fixed lines decreased in most countries from 2002 until 2007 (ITU, 2008).

References

Ba, A. 2003. *Internet, Cyberespace et Usages en Afrique*. Paris: L'Harmattan.

Benhamou, B. 2002. *Les Enjeux Politiques de l'architecture et de la Régulation de l'internet, Les Cahiers du Numérique*. Paris: Editions Hermès CNRS.

Bernard, E. 2003. Le déploiement des infrastructures Internet en Afrique de l'Ouest. Ph.D. Thesis, University of Montpellier III.

Bonjawo, J. 2002. *Internet: Une chance pour l'Afrique.* Paris, Karthala.

Brunet, P. J., Tiemtore, O., and Vettraino-Soulard, M.-C. 2002. *Les enjeux éthiques d'Internet en Afrique de l'Ouest.* CRDI-Presses de l'Université Laval.

Cheneau-Loquay, A., ed. 2000. *Enjeux des technologies de la communication en Afrique: du téléphone à Internet.* Paris: Karthala.

———. 2002. "Défis liés à l'insertion des technologies de l'information et de la communication dans les économies africaines. L'exemple d'Internet au Sénégal." In Djeflat, A. et Boidin, B., *Ajustement et technologie en Afrique*, Publisud.

———. 2004a. "Comment les NTIC sont-elles compatibles avec l'économie informelle en Afrique ?" In *L'Annuaire français des relations internationales*, Vol. 5, Paris: La Documentation française & Bruylant, 345–375.

———., ed. 2004b. *Mondialisation et technologie de la communication en Afrique.* Paris: Karthala-MSHA.

Crouzet, T. 2007. *Le Cinquième Pouvoir. Comment internet bouleverse la politique*, Bourin Editeur, Paris.

Elkyn, C. 2001. "Concurrence mondiale et NTIC : La chance de l'Afrique." In Tshimbulu N. R., ed., 2001.

Essono, L. M. O. 2004. "Langues nationales et NTIC : éclosion linguistique ou phagocytose." In Cheneau-Loquay A., ed., 2004b.

Frère, M.-S. 2000. *Presse et Démocratie en Afrique francophone. Les mots et les maux de la transition au Bénin et au Niger.* Paris: Karthala.

Frère, M.-S. and Heuzé, F. 2007. "Cultures africaines et Mondialisation." In *Les Cahiers Marxistes*, n 236, October 2007.

Fullsack, J. L. 2002. "L'UIT: la vieille dame des télécommunications dans la tourmente néo-libérale." http://www/csdptt.org.

Heuzé, F. 2007. Les NTIC en Afrique de l'Ouest, Master Thesis, Université Libre de Bruxelles, Département des Sciences de l'Information et de la Communication.

ITU. 2006. *World Telecommunication/ICT Development Report 2006: Measuring ICT for Social and Economic Development.* Geneva, 8th Edition.

ITU. 2008. *African Telecommunication / ICT indicators 2008: At a Crossroads*, Geneva.

Kiyindou, A. 2006. "La place des savoirs africains sur Internet ou penser 'la fracture numérique' par le contenu." In *Netsuds* n 2.

Koudjou-Talla, C. L. 2007. *Etude de l'Impact des Centres Aden au Cameroun*, Yaoundé.

Laborde, A. 2006. "TIC et développement humain. Analyse de discours." In *NetSuds* n 2.

Lenoble-Bart, A. 2006. "Info-riches et info-pauvres : 'fossé numérique et solidarité numérique dans la cyberpresse en Afrique.'" In *NetSuds* n 2.

Levine, V. T. 2004. *Politics in Francophone Africa*, Boulder, Lynne Rienner Publishers.

Lohento, K. 2006. "Ong et appropriation des technologies modernes de communication en Afrique." In *NetSuds* n 2.

Loiret, P. J. 2005. *L'Université virtuelle africaine. Histoire d'une mise en scène (1997–2004)*, Ph.D. Thesis, IRD, Geneva.

Mbengue, M. 2002. "Internet et enjeux culturels en Afrique." Master Thesis, EBAD, University Cheikh Anta Diop, Dakar.

Mezouaghi, M. 2004. "TIC et globalisation: enjeux industriels pour les pays en développement." In Cheneau-Loquay Annie, ed, 2004b.

Mignot-Lefèvre, Y. 2006. "L'après développement et les technologies numériques." In *Netsuds* n 2.

Pare, C. 2007. Médias et société de l'information en Afrique de l'Ouest. Enjeux, discours et appropriations, Ph.D. Thesis, University of Bordeaux III.

Perrineau, P. 2007. *Le désenchantement démocratique*, Paris, L'Aube.

Renaud, P. 2000. "Histoire de l'Internet du Nord au Sud." In Cheneau-Loquay A., ed., 2000.

Renaud, P. 2004. "Internet Nord-Sud : Fossé ou passerelle numérique," ICT Conference, IRD (November 2004). Available on www.tic.ird.fr/article.php?id_article=29

Renaud P. 2005. "La fracture numérique Nord-Sud et les moyens engagés pour la réduire," May 2005. http://www.tic.ird.fr/article.php?id_article=126.

Seck, M. T. 2000. "Insertion d'Internet dans les milieux de la recherche scientifique." In Cheneau-Loquay A., ed., 2000.

Sissouma, S. 2001. "La presse africaine face au défi des nouvelles technologies." In *Les Cahiers du journalisme* n 9, ESJ-Lille. Québec, France: University of Laval.

Tebeje, A. 2005. "L'exode des cerveaux et le renforcement des capacités de l'Afrique," site du CRDI. http://www.idrc.ca/fr/ev-71249-201-1-DO_TOPIC.html.

Tshimbulu, R. N., ed. 2001. *L'Internet, son Web et son e-mail en Afrique. Approche critique.* Paris: L'Harmattan.

Tudesq, A.J. 1998. *Espoirs et illusions. Actions positives et effets pervers des médias en Afrique subsaharienne.* Talence: MSHA.

World Bank. 2006. *Information and Communications for Development 2006: Global Trends and Policies.* April.

CHAPTER 6

Use of the Internet by NGOs to Promote Government Accountability: The Case of Egypt[1]

Khayrat Ayyad

Introduction

From its inception as an information technology, the Internet has moved on to become a tool with the capacity to influence almost all aspects of human life. These include its use as a tool for persuasion, pro-social behavior, intergroup conflict, leadership, group discussion, and decision-making (Amichai-Hamburger 2008, 544). The Internet in the last decade has brought about dramatic changes not only in how individuals carry out day-to-day communications but also in how organizations conduct their business (Ayish 2004, 381; Kirat 2007, 166).

For corporate communications, Internet technologies offer many possibilities, from text-based e-mail to transmission of images and sound, even via own Web sites (Alfonso and Miguel 2006, 267). Organizational Web sites provide challenges and opportunities for public relations (PR) practitioners by building dialogue with its different publics (Kent and Taylor 1998, 321). The web also has the ability to express the positions of different stakeholders in the communication process more equitably (Naude et al. 2004, 87).

Before the advent of the World Wide Web (WWW), advertising was the only means to send a controlled message to a mass audience through a mass medium. WWW can be considered the "first controlled mass medium" (White and Raman 2000, 405) for organizations as it allows messages to flow directly to targeted publics without the gatekeeping function of journalists and editors. As a result, the use of the web could have particular advantages for Non-governmental Organizations (NGOs) that cannot afford paid advertising (Kent et al. 2003, 63). Many researchers agree that the growth of the Internet is playing a vital role in the emersion, evolution, and development of civil society organizations (CSOs). It is revolutionizing the NGOs by facilitating the formation of networks and coalitions and even transnational social movements (Clark and Themudo 2006, 50).

Through the use of their Web sites, NGOs can empower themselves, be more widely heard, be more visible, and can interact with potential sponsors as well as their publics on a low budget. With limited resources, a Web site gives organizations, such as the NGOs, the chance to participate in discussions on public policy issues, increase the effectiveness of their communications, build links with their publics, and strengthen their opportunities to challenge the practices of large corporations and governments (Naude et. al. 2004, 87; Ryan 2003, 335).

The Internet also gives ample opportunity to members of NGOs to exchange information, thereby bringing members to a relational space together. It is a relevant channel for NGOs to communicate their messages to gain public support for confronting issues (Amichai-Hamburger 2008, 544; Kang & Norton 2004, 279).

The Internet can fulfil four main goals for NGOs: borderless communication, inclusive communication, continuous communication, and dialogue (Ranchhod et al. 2002, 6). Through their Web sites, NGOs can introduce local issues to the global arena. This facility gives NGOs the capability to be transnational and transcultural networks. The Internet also facilitates inclusive communication through the reach it has on the public sphere, both internal and external, which help these organizations become news sources. The WWW enables NGOs to build issues or cases concerning local and national matters through continuous and sustained communication. Communication is a process of exchange, and the unidirectional model is becoming obsolete. Through the Internet, NGOs can perceive and understand the public interest and thus enhance dialogue and interactive communication (Clark & Themudo 2006, 50; Ranchhod et al. 2002, 6).

Internet and NGOs in Egypt

Since the 1980s, Egypt has been moving full speed ahead in the technological arena. Egypt's telecom sector is performing consistently well, with most subsectors open to competition. Around 3,000 new fixed lines are installed in the country every day, a trend that has helped to reduce the waiting list by around 90 percent in recent years. The incumbent telco, Telecom Egypt, is highly profitable and was partially privatized through an initial public offering at the end of 2005. The end of its fixed-line monopoly in 2006 opened up new opportunities for competitive service providers, as did the award of a third mobile license. Efforts are underway to roll out Next Generation Networks (NGN) offering converged IP-based voice and data services (MiniWatts 2007).

In 2000, only 0.7 percent of Egypt's nearly 67 million people were Internet users. Due to the successful implementation of a free Internet strategy in 2002, Egypt has the largest Internet market in Africa, with 8.8 million users, 8.3 percent of the population, in 2007—up from more than five million in early 2006. The sector is highly competitive with around 300 Internet and data service providers. A broadband initiative launched by the government in 2003 increased the number of broadband connections tenfold from 2003 to 2006 and has brought 24Mb/s ADSL2+ access to residential households (MiniWatts 2007). The government provides free Internet access to governmental agencies, NGOs, and other organizations. Furthermore, under its e-Access initiative, the government and a number of private organizations have organized a number of programs that include free PCs for citizens and free Internet services nationwide to promote computer literacy (Ford 2007, 302). However, Internet penetration is still relatively low and the vast majority of its users are located in urban areas.

As for the NGOs, this sector in Egypt is very large, with approximately 15,000 NGOs currently operating. They are predominantly small community-based organizations that concentrate on service delivery and social assistance. Roughly 11,000 of these NGOs have been actively working throughout the country to provide health care, education, job training, child care, elder care, welfare, legal assistance, human rights, water, irrigation, environmental, and other social and economic services to a largely poor population. Community development associations, Islamic and Christian charitable groups, feminist organizations, student groups, and (more recently) capitalist associations are active in satisfying their own publics and the needs of their own communities. The past decade has seen a rise in "advocacy" NGOs, and there is an expectation that they would push forward on social and political change in a way that weak political parties and professional syndicates and unions suffering from government interference were not doing; however, they have had a limited impact (Ibrahim et al. 2003; Sullivan 2000).

It can be argued that civil society, including NGOs, in Egypt is weak compared to the strength of the state. Virtually all participants in, and observers of, NGO activities in Egypt recognize that these organizations are far from being independent of the government and many in fact are creations of that government. Over the past year, the government has tightened its control over civil society institutions and restricted freedom of speech and expression. Government interference seriously limits the autonomy of civil society actors, including NGOs, professional syndicates, and intellectuals (Sullivan 2000).

In addition to difficulties stemming from government interference, NGOs are facing many obstacles, such as inadequate resources, weak organizational setup, lack of routine external audits, and absence of strict internal rules and regulations. Egyptian NGOs operate in a precarious financial situation. They are heavily dependent on foreign funding, which is problematic for several reasons. Receiving funding from external sources can hurt their image by opening them up to criticism from the government, that is, they are facilitating foreign interference (Ibrahim et al. 2003).

One of the main problems confronting NGOs in Egypt is access to mass media and access to information, because of limited resources. The media in Egypt are strictly regulated by a combination of structural arrangements, whereby the government monopolizes the press and broadcasting, and legal controls prevent journalists from reporting freely on sensitive domestic issues or deviating from official foreign policy when reporting on international affairs. The content of the mass media always reflects the view of the political elite and follows the laid down national policy (Ibrahim et al. 2003; Ayyad 2001).

Therefore the Internet becomes the most important medium for NGOs to reflect their views about socioeconomic issues and current affairs. The Internet enables NGOs in Egypt to overcome financial limitation, difficulties of accessing information, and censorship in their expression of concerns. There is no data on how many NGOs have a Web site, but most of them have developed their own Arabic language homepages, and a significant number maintain bilingual Web sites—Arabic and English. These Web sites are used by NGOs to address many socioeconomic issues as well as human rights.

Corruption is one of the issues CSOs and NGOs in Egypt attempt to tackle using ICTs such as the Internet. Researchers agree that corruption poses a serious development challenge, seriously affecting the economy. In the political realm, it undermines democracy and good governance by flouting formal processes. Corruption in elections and in legislative bodies reduces accountability and distorts representation in policymaking, corruption in the judiciary compromises the rule of law, and corruption in public administration results in the unfair provision of services (Hopkins 2002, 574). Corruption in Egypt is a serious problem, as evidenced in a report by the Egyptian coalition opposition group Kefaya (2008) and the corruption index compiled by Transparency International (TI 2007).

Research on Organizations Communicating via the Web

According to Kent et al. (2003, 63), research about the Internet and organizational communication fall into two main categories: characteristics of the design of Web sites for providing information to the public, and attitudes and expectations about the web as an organizational communication tool. As this study falls in the first category, the following discussion focuses on how the Internet is used by different organizations to fulfill their objectives.

Taylor, Kent, and White (2001, 63) conducted a study of 500 organizations and concluded that organizations should design their Web sites strategically to meet the information needs of their stakeholders. Ayish (2004, 381) analyzed Web sites of twenty public and private organizations in the United Arab Emirates (UAE) to examine patterns of their usage of the Internet as a communication tool in terms of availability of some features. He concluded that these organizations use the Internet to disseminate information about their identity and activities. The organizations are also committed to perform the two-way symmetrical model of communication with their publics, which is one of the best uses of Web sites. The model, introduced by Grunig and Hunt in the late 1980s, is considered a useful theoretical framework to study PR in an interactive, new media context (Naude et al. 2004, 87).

In a study of twenty-four PR departments in public and private organizations in the UAE, Kirat (2007, 166) found out that they did not effectively promote online media relations. They failed to use facilities that enhance media relations—such as virtual tour, film and videos, newsrooms, and speech archives. Kang and Norton (2004, 279) analyzed the Web sites of 100 nonprofit organizations in the United States to explore how they use the Internet to accomplish organizational goals. They concluded that most sampled organizations did not utilize the benefits of the web to attract more visitors and that they need to develop a better quality design for their Web sites.

One of the most frequent functions of a Web site is to provide information about the organization to its stakeholders. Organization Web sites incorporate various types of content (Ryan 2003, 335), which varies from one organization to another. The content of an activist organizations' Web site usually concentrates on public issues and share common goals such as cleaner environment, equal rights, improved health care, transparency, accountability, and human rights (Kent et al. 2003, 63; Clark and Themudo 2006, 50).

Several researchers have stressed the possibility for dialogue and long-term relationships through the use of the web. Dialogue theory suggests that organizations must be willing to interact with the public in honest and ethical ways in order to create effective organization-public communication channels (Ryan 2003, 35). Taylor et al. (2001, 263) examined the use of the Internet by activist groups and found that activist organizations do meet the informational needs of members of the public but have yet to facilitate creative dialogue between organizations and their stakeholders.

Kent et al. (2003, 63) analyzed the Web sites of activist organizations to examine the extent to which these Web sites could be used for interactive communications with stakeholders. They concluded that the activist organizations employed poor dialogic communication and exhibited little commitment to building relationships with the interested public. They suggested that if organizations want to use their Web sites to build relations with the public, certain design features are necessary. The more the organization depends upon its public members for accomplishing its mission, the more it should employ dialogic features into its Web sites design.

Research Questions

As this study aims at exploring how the NGOs in Egypt use the Internet to promote government accountability concerning socioeconomic issues, it answers the following two questions:

- What kinds of issues relating to government accountability and socioeconomic development are presented on the Web sites of NGOs in Egypt? To what extent does the type of NGOs affect this presentation?
- To what extent do these Web sites employ interactive communication to promote government accountability?

Methodology

There are about 15,000 NGOs formally registered with the Ministry of Social Affairs (MOSA) in Egypt. Figures provided by MOSA indicate that about 84.1 percent of the registered NGOs work in areas of social aid, local community development, and cultural, scientific, and religious services. A significant number are classified as advocacy organizations. It is difficult to have a definite classification for NGOs in Egypt. Some organizations are primarily registered with MOSA as social aid organizations or as service delivery organizations, yet they sometimes operate as activist or advocacy groups (Ibrahim et al. 2003). As a result, two main categories of NGOs in Egypt can be determined. The first comprises

NGOs that concentrate on political issues and human rights. They will be classified as Advocacy NGOs. The second category includes NGOs that focus on social and economic services. These organizations will be classified as Non-advocacy NGOs.

Among a sample of 102 NGOs that was selected for the study, 49 comprised Advocacy NGOs and 53 were Non-advocacy NGOs. All of these NGOs have Web sites on the Internet. Some of these Web sites are bilingual (Arabic & English), while others are only in Arabic.

A coding sheet was designed to carry out the analysis. The coding sheet contains two main categories: content, and presentation. Content categories include many subcategories covering aspects and issues related to obstacles of socioeconomic development and accountability. The presentation of content categories includes subcategories covering facilities of disseminating information and data, getting feedback, and performing dialogue and interactive communication. Categories of the coding sheet are designed to reflect how serious each issue affects socioeconomic development. A scale of five degrees, 1 to 5, has been developed to grade the organizations—where 1 means no effect and 5 means too serious. Descriptive statistical analysis of one-sample t-test and two-sample t-test is applied to examine variables of the study. The analysis process was carried out in February 2008.

Findings

As explained earlier, corruption poses a serious challenge for all aspects and dimensions of development in Egypt. According to reports published by the Egyptian Organization of Human Rights (www.eohr.org), the most serious corruption problems facing development in Egypt concern information deficits and human rights. As explained above, many NGOs in Egypt are using their Web sites to convince the Egyptian government that development programs will be useless without protecting human rights and securing freedom of expression and access to information. Many foreign institutions also have asked the Egyptian government to improve its record of human rights. In a resolution adopted on Egypt in January 2008, the European Parliament called on the Egyptian government to implement a range of measures to improve democracy and human rights, such as ending the harassment of journalists and human rights defenders, lifting the state of emergency, investigating suspected cases of torture, and guaranteeing the independence of the judiciary (Resolutions of European parliament). In this respect, the following section discusses how the Web sites of NGOs deal with information deficits and human rights as problems facing socioeconomic development in Egypt.

Internet and Information Deficits

The following table (table 6.1) summarizes the use of the Internet by NGOs to reveal how the information deficit seriously affects development.

The figures in table 6.1 show that almost all information deficits have serious and negative effects on socioeconomic development. The mean of these deficits exceed three, which is the test value of the one-sample t-test. The most serious obstacles that poses serious burden to development are lack of government transparency, lack of freedom of the press, and lack of freedom of expression and the right to demonstrate. The analyzed Web sites of NGOs published detailed reports explaining how serious these issues are. The government is losing its credibility among people because of lack of transparency. With such serious restrictions over the press, it is difficult for journalists to monitor policies and activities of the government.

In a situation where the main figures in national newspapers and the Radio and TV Union are accused of corrupt activities, it may be argued that the Internet represents the most suitable medium to experience free press and expression in Egypt.

To what extent does the nature of NGOs affect their dealing with issues related to information deficits? The following table (table 6.2) illustrates how Web sites of Advocacy and Non-advocacy NGOs present information deficits as serious obstacles to socioeconomic development.

Figures in the above table indicate that there is a significant difference between Web sites of Advocacy and Non-advocacy NGOs in dealing with issues relating to information deficits. While Web sites of Advocacy NGOs show that all these issues pose too serious a burden on socioeconomic development, it can be noted that Web sites of Non-advocacy

Table 6.1 Burden of issues related to information deficits

One-Sample t-test*

Issues related to information deficits	*Mean***	*Standard Deviation*	*Sig.****
Lack of government transparency	3.23	1.58	0.15
Lack of freedom of press	3.16	1.67	0.35
Lack of freedom of speech and demonstration	3.11	1.56	0.49
Lack of freedom of information legislations	3.09	1.64	0.59
Lack of measurements of corruption	2.19	1.45	0.00

Notes:
* N (sample size) = 102
** Test value = 3
*** Confidence interval = 95 percent
Sig. = Significance level

Table 6.2 Type of NGOs and issues related to information deficits

Two-Sample t-test

Issues related to information deficits	*Means in Advocacy NGOs**	*Means in Non-advocacy NGOs***	*t*	*Sig.****
Lack of government transparency	4.35	2.19	9.44	0.00
Lack of freedom of information legislations	4.39	1.89	11.98	0.00
Lack of freedom of press	4.53	1.89	13.11	0.00
Lack of freedom of speech and demonstration	4.27	2.04	10.28	0.00
Lack of measurements of corruption	2.88	1.55	5.20	0.00

Notes:
* N (sample size) = 49
** N (sample size) = 53
*** Confidence interval = 95 percent

NGOs consider these issues not to have much relevance or effect. As Advocacy NGOs are mainly interested in discussing issues relating to all dimensions of national development, they use their Web sites intensively to communicate with the public and to reflect their views about the relations between information deficits and socioeconomic development. The content of these Web sites repeatedly considered transparency, freedom of the press, and freedom of expression as important requirements of socioeconomic development.

Internet and Government Accountability

In a society where the traditional media are used to support government and justify its activities, the Internet is seen as an important alternative to monitor these activities. The following table (table 6.3) shows issues related to government accountability and to what extent these issues represent serious burdens to socioeconomic development.

It can be seen from the following table that the most serious issue confronting socioeconomic development is the lack of civil society. Web sites of NGOs reflect the situation in Egypt where the government usually tightens its control over civil society institutions and restricts freedom of speech and expression. In the last few years the NGOs have used the Internet to build many cases against the government. Most of these cases related to violation of human rights by the police in prisons. Many police officers were taken to court and found guilty of violating the rights of civilians. Many NGO Web sites have documentary films and testimonies

Table 6.3 Burden of issues related to Government accountability

One-Sample t-test*

Issues related to monitoring and accountability	*Mean***	*Std. Deviation*	*Sig.****
Lack of civil society and NGOs	3.39	1.61	0.02
Democracy dysfunction	3.36	1.62	0.03
Weak rule of law	3.36	1.58	0.00
Police violation of human rights	3.25	1.69	0.13
Rational ignorance of votes accounting	3.21	1.60	0.20
Weak judicial independence	3.13	1.52	0.40

Notes:
* N (sample size) = 102
** Test value = 3
*** Confidence interval = 95 percent
Sig. = Significance level

revealing violent acts perpetrated by police officers against civilians all over Egypt.

Two other serious issues influencing socioeconomic development are a disfunctioning democracy and the absence or the ignorance of the rule of law. NGOs argue through their Web sites that the Law of Emergency imposed in 1981 is the main reason for the deterioration in developmental situation in Egypt. It allows for arbitrary detention of individuals for an indefinite period without granting a fair trial. According to Human Rights Watch report, this law reflects a punitive regime that is always suspect.

The Web site of an advocacy organization, Human Rights Centre on Prisoners Help, mentioned that the number of long-sentence prisoners in 2004 who faced no charges reached about 15,000 persons. The Centre issued a list of about sixty-five lawyers who were detained for about sixteen years with no charge or trial. NGOs also argued that Egypt cannot move forward without ending the state of emergency.

Figures in the above table (table 6.3) also indicate that one of the main and serious issues relating to political development, and which affect credibility of government, is vote counting during general election. All the Advocacy NGO Web sites contain archived and documented reports concerning various general elections, which brought the same party, National Democratic Party (NDP), to the office in the three decades till 2008. Many NGO reports indicate that these kinds of elections destroy credibility of the political regime and weaken legislative authority.

The following table (table 6.4) shows the relationship between the type of NGO and issues related to government accountability.

Table 6.4 Type of NGOs and issues related to information deficits

Two-Sample t-test

Issues related to accountability	*Means in Advocacy NGOs**	*Means in Non-advocacy NGOs***	*t*	*Sig.****
Democracy dysfunctional	4.53	2.28	9.76	0.00
Lack of civil society & NGOs	4.51	2.36	9.03	0.00
Rational ignorance of votes accounting	4.39	2.11	10.19	0.00
Weak rule of law	4.51	2.58	7.75	0.00
Weak judicial independence	4.22	2.11	9.73	0.00
Police violation of human rights	4.31	2.28	7.50	0.00

Notes:
* N (sample size) = 49
** N (sample size) = 53
*** Confidence interval = 95 percent
Sig. = Significance level

The above table indicates that the NGO type significantly influences its views toward issues relating to government accountability. Web sites of Advocacy NGOs conclude that all of these issues have too serious effects on socioeconomic development. Statistical analysis shows that the mean of each issue exceeds four. On the other hand, it seems that Web sites of Non-advocacy NGOs are not interested in discussing issues relating to government accountability. The mean of each issue accounted for less than three, which was the test value.

Presence of Feedback and Dialogue

Researchers agree that one of the most important capabilities of the Internet is to perform and develop a two-way symmetrical communication approach. They also stress the possibility for dialogue and long-term relationships through the use of the web (Kent and Taylor 2002). Table 6.5 summarizes patterns of facilitating dialogue and easing feedback used by Web sites of NGOs.

It can be seen from table 6.5 that facilities for getting feedback and performing interactive communication with the public are not important. Figures in the above table show that the means obtained for all the facilities of feedback are less than the minimum value of the used scale. In analyzing Web sites of NGOs, it was noted that only two sites out of the analyzed sample had some links to interactive communication. Three sites only designed and conducted some kind of survey on specific issues. It was also noted that the most frequent way of connecting NGOs to

Table 6.5 Presence of feedback and dialogue

One-Sample t-test*

Feedback & Dialogue	*Mean***	*Std. Deviation*	*Sig.****
Email	0.97	0.17	0.00
Opportunities to vote on an issue	0.20	0.40	0.00
Surveys	0.10	0.30	0.00
Links to submit complaints	0.40	0.49	0.00
Links to submit suggestions	0.22	0.41	0.00
Links to interactive communication	0.07	0.25	0.00

Notes:
* N (sample size) = 102
** Test value = 3
*** Confidence interval = 95 percent
Sig. = Significance level

Table 6.6 Type of NGOs and feedback facilities

Two-Sample t-test

Feedback and Dialogue	*Means in Advocacy NGOs**	*Means in Non-advocacy NGOs***	*t*	*Sig.****
E-mail	1.00	0.94	1.70	0.09
Opportunities to vote on an issue	0.41	0.00	5.99	0.00
Surveys	0.20	0.00	3.65	0.00
Links to submit complains	0.55	0.26	3.06	0.00
Links to submit suggestions	0.37	0.08	3.79	0.00
Links to interactive communication	0.06	0.08	−0.28	0.78

Notes:
* N (sample size) = 49
** N (sample size) = 53
*** Confidence interval = 95 percent
Sig. = Significance level

their public is via e-mail. It seems that all NGOs rely on this facility to obtain feedback from the public. Many Web sites indicated that filmed and documented materials were mailed to the site by ordinary people.

The above table (table 6.6) shows the presence of feedback facilities on sites of Advocacy and Non-advocacy NGOs.

Figures in the above table show that while the NGO type has no significant effect on using facilities of feedback, it can be noted that sites of Advocacy NGOs employ some facilities more than Non-advocacy NGOs. These facilities provide opportunities to vote on current issues and also to submit complaints and suggestions.

Discussion

Returning to the first question of the study, it can be argued that NGOs in Egypt have made use of their Web sites to discuss serious issues and obstacles concerning socioeconomic development. These issues relate mainly to negative effects of corruption and human rights. The contents of these sites concentrate on problems facing the media environment and freedom of the press and freedom of expression. Issues relating to transforming Egypt toward a civilized and democratic society have also received a great deal of attention.

For question two, it may be argued that Web sites of NGOs lack the facilities to provide an effective feedback to the organizations' members and the public. The majority of Web sites also have poor facilities to conduct online survey or even to ask closed questions regarding current affairs. There is also no evidence of any facilities for interactive communication or ways of encouraging dialogue with the organizations' public. However, it can be argued that the Internet is still the most effective medium to promote government accountability and monitor its activities, particularly when there are major restrictions imposed on the practice of both print and broadcast media in Egypt.

There are plenty of evidences to prove that the Internet has an important role in a country such as Egypt. For example, as a result of rising food prices in Egypt, a call for a strike on April 6, 2008 was resorted to by a group of Egyptian youths. The group, which organized themselves through Facebook networking, succeeded in mobilizing more than eighty thousands supporters against the government. Hundreds of thousands of Egyptians joined the strike. The Facebook group soon came under intensive pressure by the state-owned media, and was accused of being a serious threat to state stability. This case indicates that the Internet has become not only an effective tool of communication but also a tool wielding significant political force.

Note

1. A version of this chapter was submitted as a paper at the Conference, "Media and Development: Local and Global Initiatives," the Communication and Media Research Institute (CAMRI), University of Westminster, London, March 27–28, 2008.

References

Alfonso, G. and R. Miguel. 2006. "Trends in Online Media Relations: Web-based Corporate Press Room in Leading International Companies." *Public Relations Review* 32: 267–275.

Amichai-Hamburger, Y. 2008. "Potential and Promise of Online Volunteering." *Computers in Human Behavior* 24: 544–562.

Ayish, M. 2005. "Virtual Public Relations in the United Arab Emirates: A Case Study of 20 UAE Organizations' Use of the Internet." *Public Relations Review* 31: 381–388.

Ayyad, K. 2001. The Developmental Dimensions of the Egyptian Media Role in the Peace Process. PhD Thesis. Leicester: University of Leicester.

Clark, J. and N. Themudo. 2006. "Linking the Web and the Street: Internet-based 'Dotcauses' and the 'Anti-Globalization' Movement." *World Development* 34: 50–74.

EOHR. 2008. Egyptian Organization of Human Rights Reports 2008. http://www.eohr.org (accessed January 13, 2008).

Ford, D. 2007. "Technologizing Africa: On the Bumpy Information Highway." *Computers and Composition* 24: 302–316.

Hopkins, J. 2002. "States, Markets and Corruption: A Review of some Recent Literature." *Review of International Political Economy* 9: 574–590.

Ibrahim, Na., A. Lachant, and L. Nahas. 2003. NGOs as Civil Society Actors on Media Policy Change in Egypt: Capacity Building within a Contextual Framework. Report Produced for Stanhope Centre for Communications Policy Research. London: London School of Economics, Development Studies Institute.

Kang, S. and H. Norton. 2004. "Non-Profit Organizations' Use of the World Wide Web: Are They Sufficiently Fulfilling Organizational Goals?" *Public Relations Review* 30: 279–284.

Kefaya. 2008. Kefaya (enough) to Cleanse Egypt, Judicial and documentary file 2008. http://kefaya.org/index.htm (accessed January 4, 2008).

Kent, M. and M. Taylor. 1998. "Building Dialogic Relationships through the World Wide Web." *Public Relations Review* 24: 321–334.

Kent, M. and M. Taylor 2002. "Toward a Dialogic Theory of Public Relations." *Public Relations Review* 28: 21–37.

Kent, M., M. Taylor, and W. White. 2003. "The Relationship between Web Site Design and Organizational Responsiveness to Stakeholders." *Public Relations Review* 29: 63–77.

Kirat, M. 2007. "Promoting Online Media Relations: Public Relations Department Use of Internet in the UAE." *Public Relations Review* 33: 166–174.

MiniWatts. 2007. Egypt Internet Usage and Marketing Report 2007. Telecommunications Industry Reports. http://www.internetworldstats.com/af/eg.htm (accessed January 12, 2008).

Naudé, A., J. Froneman, and R. Atwood. 2004. "The Use of the Internet by Ten South African Non-governmental Organizations—A Public Relations perspective." *Public Relations Review* 30: 87–94.

Ranchhod, A, C. Gurau, and J. Lace. 2002. "On-line Messages: Developing an Integrated Communications Model for Biotechnology Companies." *Qualitative Market Research* 5: 6–18.

Resolutions of European Parliament. http://www.europarl.europa.eu/sides/getDoc.do?language=EN&type

Ryan, M. 2003. "Public Relations and the Web: Organizational Problems, Gender and Institution Type." *Public Relations Review* 29: 335–349.

Sullivan, D. 2000. *NGOs and Development in the Arab World: The Critical Importance of a Strong Partnership between Government and Civil Society.* Published on the Web with Permission of Ibn Khaldun Centre, Cairo. http://www.mideastinfo.com/arabngo.htm (accessed January 18, 2008).

Taylor, M, M. Kent, and W. White. 2001. "How Activist Organizations are Using the Internet to Build Relationships." *Public Relations Review* 27: 263–284.

TI. 2007. Corruption Perceptions Index, Regional Highlights: Africa 2007. Transparency International (TI). http://www.transparency.org/policy_research/surveys_indices/cpi/2007 (accessed August 1, 2008).

White, C. and N. Raman. 2000. "The World Wide Web as a Public Relations Medium: The Use of Research, Planning, and evaluation in Web Site Development." *Public Relations Review* 25: 405–419.

CHAPTER 7

ICTization beyond Urban Male Elites: Issues of Gender Equality and Empowerment

Kutoma J. Wakunuma-Zojer and Patricia K. Litho

Introduction

In the past few decades, the information and communication technology (ICT) "revolution" has promised a variety of benefits to society, including material progress and subsequent empowerment for women and men in the developing world. However, there are parallel discourses that question the promised benefits of ICTs and the extent to which the rhetoric of "empowerment," particularly for women in traditional and rural settings, is being transformed into practice on the ground. This chapter analyses the efficacy of ICT "empowerment" projects targeted at women, especially in rural Africa. The chapter discusses two ICTs, computer-based Internet and cellular telephony, assessing mainly their political—as well as social and economic—impacts with regard to "empowerment" of rural women in Uganda and Zambia, because of the similarities in respect to women's experiences of ICT in as far as empowerment is concerned in both these countries.

Overview

Most ICT projects, often telecenters targeted at women in developing countries, usually privilege PC-based Internet (hereafter simply referred to as "Internet") at the expense of other ICTs such as basic voice and text message, and cell phone applications (hereafter simply referred to as "cell phone"). The dominance of the Internet globally is not surprising, as it allows a convergence of other technologies, and it is "a crucial channel for the diffusion of essential information in those places where poverty implies not only lack of economic means but also lack of accurate information" (Murelli and W'O Okot-Uma 2002, 3). The Internet has networked society and the world in unprecedented ways (Roy 2005; Cairncross 1998; Castells 1997). These realities play a part in the bias of ICT projects in favor of the Internet. However, this position is at odds with what prevails in most developing countries, such as Uganda and Zambia, when it comes to the diffusion and adoption of ICTs, because cell phone is more preferred and ubiquitous due to its scalability, impact, importance, and relevance.

As Ling (2004, 16–17) illustrates, the cell phone "is, in itself, more accessible than the PC/Internet. The technology is relatively inexpensive and widely available." Due to an underdeveloped telecommunication infrastructure, the growth of cellular telephony has been more rapid (Burgess 2004). Even in Africa, unlike the Internet, the cell phone is no longer a luxury exclusive to "the wealthy and educated, at least, to leapfrog an entire generation of technological progress" (Burgess 2004, 34–35). Horst and Miller (2006, 11) thus conclude that "the cell phone mushrooms up from inside mud-brick shacks and under corrugated iron sheet roofing to become an insistent and active presence."

In picking up on the strand of "leapfrogging" technology, Butler (2005) alludes to mobile Internet via cell phone, without the need for an expensive computer. However, in countries like Uganda and Zambia, especially for the poor, most people can only afford a very basic phone, only for voice calls and possibly short messaging service (SMS). Of concern to such users are factors such as handset cost and airtime affordability. In essence, it has been an oversight to privilege the Internet over the cell phone particularly in ICT projects such as telecenters.

This chapter looks at gender differentials in the access and use of the ICTs in question and discusses some of the policy issues in Uganda and Zambia—examining how these reflect on the aspects of gender equality. We ground our discussion in democratic theory and social constructivist perspectives in order to examine the efficacy of ICTs for women's empowerment and gender equality beyond urban male elites. To this effect the chapter also endeavors to conceptualize empowerment.

Theoretical Perspective

Democratic theory is about individual rights to participate in decision-making processes and how they can be governed by their leaders. It is concerned with the issues of accountability and individual autonomy (See Rawl 1985; Mulhall and Swift 1992). In terms of women's rights, gender equality, and empowerment, democratic theories are particularly useful for teasing out the extent to which women have been able to exercise their democratic rights as a result of exposure to ICTs such as the Internet and cellular telephony. We argue for a move beyond the telecenters model since the issues of gender inequality and empowerment in as far as ICTs are concerned go beyond access to ICTs within such a narrow confinement and rather include experiences from the wider environment that women exist in. Telecenters, or such similar arenas, intended for women will most likely alienate them further from the mainstream decision-making machinery as they remain confined in the periphery. Although such environments can be applauded because their intention is to get women involved in using ICTs, often the expected outcomes do not go further to incorporate them into the mainstream; they remain the "other"—thus reinforcing gender inequalities.

Constructivist perspectives (Heap et al. 1995) take into consideration the fact that oftentimes technologies are interpreted differently by different users. This may be based on their gender as well as sociocultural, political, economic, and historical contexts as opposed to the hegemonic posit[illegible] painted, which seem to represent a single perspec[illegible]. Using constructivist perspectives in considering [illegible] between ICTs, women's empowerment, and gender [illegible] to recognize that technology adoption and use [illegible] power relations within social groups and the dif[illegible] interpret a technology. Faulkner (2000) observes [illegible] in the use of and access to technology can also [illegible] artifacts. She categorizes the "genderedness" [illegible] gender in technology" and "gender of technol[illegible]ated with artifacts designed with either men or women in mind, with gender already embodied, constructed or reinforced in the design process. In the latter, certain home-based technologies like the washing machine, the microwave (also looked at by Cockburn and Ormrod 1993), or the video cassette recorder, though not labeled as either intended for men or women, are often associated with one or the other gender. As such, various actors may shape the direction of an artifact in their capacity as powerful individuals or in their political leadership positions. Individuals or institutions as legal entities may also have the potential to influence the adoption of a specific artifact because of their strong financial or technical positions. It is also important to note that the less powerful actors, who happen to be the majority of users, may be able to completely change the direction of an artifact because they constitute a community of consumers that is able to influence market forces to re-adapt to their demands, which is evident in the uptake of cell phones over the Internet. As a result, appropriation of a technology may differ geographically, regionally, locally, and according to one's social and economic needs, including one's gender.

In reality, the choice to construct may not be as straight forward as it may appear to be, especially when that choice is taken out of the user's hands, either through lack of consultation or feasibility study. Policymakers and development agencies tend to determine the course of action top-down, usually without consideration of target users. A constructivist approach would allow the intended beneficiaries, usually women in this case, to be able to engage—individually and collectively—with a technology to provide meaningful outcomes and a sense of ownership. The result would be empowerment.

Using empirical evidence from Uganda and Zambia, we challenge the often hegemonic and preconceived ways of looking at the potential of ICTs for women's empowerment and gender equality, which ignore the fact that experiences are context-specific.

Cell Phones versus the Internet

Internet and cell phone use is more prevalent in urban than rural settings, especially in the capital cities of both countries. For Zambia, prevalence is higher especially along the rail line where the infrastructure is better than in rural areas. While Internet activity is still very much centered in major cities, the cell phone industry has seen the creation and penetration of small-scale businesses such as cell phone repairs, airtime sale, and cell phone accessories in both urban and rural areas.

Additionally, although there is shared use of the two ICTs, shared cell phone usage seems to be on a greater scale. For example, whereas shared use of the Internet often occur in Internet cafés as well as at places of work or higher learning institutions, shared cell phone usage occurs among family and friends, through its use as a public pay phone, as well as in community centers. Street vendors who offer their cell phones as public pay phones penetrate places where the Internet often does not.

However, it must be noted that in both countries there are no official statistics that show shared use of these technologies. Our data is based on field experience and observation as well interviews with users. All the same, statistics from the Uganda Communications Commission (UCC 2006) and the ITU (2008) show the prominence of cell phones in Uganda and Zambia in the last few years. Despite this growth in ICT usage, rural areas remain disadvantaged due to infrastructural issues. While 72 percent of Uganda's connectivity is based in the urban areas, just 28 percent is in rural areas where the ratio of fixed lines is estimated at 10,000 people per line (UCC 2006).

In the case of Zambia and its Internet use, there were around 50,000 users in 2004 and this rose to about 168, 000 in 2006 (CAZ 2004, 2006). Despite such continued growth, the number is relatively very small given the country's population of about 11 million. The number of cell phone service consumers grew within a year by over 100 percent in 2005 to reach 940,000, compared to only 91,000 fixed line subscribers (CAZ 2004, 2006). Adequate and up-to-date gender disaggregated data are hardly available, but results of a national survey of 2004 by the Central Statistics Office (CSO) of Zambia showed that 2 percent of male-headed households had landline telephones while only 1 percent of female-headed households had landlines. In the case of cell phones, 5 percent male-headed households owned cell phones compared to 3 percent of female-headed households. In the case of computers, only 1 percent of male-headed households had computers while there were no statistics showing ownership of computers in female-headed households (CSO 2004). However, the statistics do not make reference to Internet use. In all this there was very little penetration, particularly of computer ownership, in rural areas.

What these statistics show, although without adequate gender disaggregated data, is first the minimal penetration of Internet connection compared to cell phone usage both in the rural and urban areas. Second, evidence of a gendered divide in terms of ICT ownership, and last, the fact that it is the cell phone, even when landline statistics are factored in, that people are using the most. This clearly poses a challenge as far as ICT access and use are concerned for those in the rural areas, particularly women. This is not least because rural areas are the most deprived in terms of infrastructure but because there are more women in rural areas when compared to men. In Zambia, for example, the rural setting accounts for 65 percent of the population, of which more than half are women. Of Zambia's 2 million households, 66 percent are rural and 68 percent of these are female-headed (CSO 2005). It is perhaps due to these considerations that one of the ways of ensuring that women in rural African countries such as Uganda and Zambia have access to ICTs is through the introduction of gender-based community centers in order to maximize access and usage.

Theorizing the ICTization of Women's Empowerment and Gender Equality: Case Studies from Zambia and Uganda

The Case Studies

The case studies used here include two gender-based ICT community projects, the Kalomo Bwacha Women's ICT Club in Zambia and the Council for Economic Empowerment of Women in Africa-Uganda (CEEWA-U) ICT Project in Uganda.

Kalomo Bwacha began as an enterprise club in 2000, comprising a group of ten women in Kalomo town in the southern Province. The objective was to support marginalized and poor women in the villages. The group changed its name to Kalomo Bwacha Women's ICT Club in October 2004, following the addition of an ICT component with support from the Netherlands' Institute for Communication and Development (IICD). Support came in the form of two computers, a mobile phone, a scanner, a printer, a digital camera, a landline phone, and a dial-up Internet connection. The main aim was to use ICTs to empower women in a variety of ways, especially economically, through the sale of services such as Internet browsing and e-mails (when there was connection), telephone services, typing, printing, scanning, and desktop publishing (creation of calendars, brochures, and cards with the use of digital photography).

CEEWA-U, an NGO comprising professional women from diverse backgrounds, set its mission to promote economic empowerment of women through advocacy, training, research, and documentation. CEEWA-U's philosophy is that "...economic strength gives women a bigger bargaining power" (CEEWA-U 2005). The ICT component of CEEWA was started in 1999 to enable women entrepreneurs[1] and women organizations that promote enterprise development to exploit ICTs for economic empowerment. The project was started with support from the International Development Research Centre (IDRC), and in 2004 it received support from the Humanist Institute for Cooperation with Developing Countries (Hivos).

Under this project, ICT facilities were provided through four telecenters at Buwama, Nabweru, Kampala, and Mukono. The facilities provided at the telecenters included computer access, Internet and e-mail services, scanning, telephone, binding, printing, and faxing facilities. Training was also offered to the women on Internet surfing, market research, business management, and book keeping, and other skills. The project also created a Web site, Women's Information Resource Electronic Service or WIRES (www.ceewauganda.org). This database-driven Web site offered information on best practices in agriculture, business skills, market prices, and financial support information. The project encouraged women in small or medium-sized enterprises (SMEs) to take advantage of the opportunities that ICTs offer. Information is also repackaged in a local language for the women.

These projects provide a basis for examining "empowerment" as it relates to women's political participation and gender equity with regard to ICTs. It is to this that we now turn.

Empowerment

A gendered divide remains in terms of access and use of ICTs in developing countries. Wajcman (1991) has argued that technology has gendered aspects and is normally biased toward men, such that it is itself often seen as "masculine." The use of ICTs can to a large extent be a catalyst to overcome this and bring about political, social, and economic empowerment for women. The word "empowerment" is often used to emphasize the potential advantages that may result through the use of ICTs, particularly for women in developing countries—bearing in mind that women are often the poorest in societies such as Uganda and Zambia. Here is how the Zambian government, in its National ICT Policy (2006), uses the term empowerment in reference to women's development:

> it is envisaged that ICTs can be harnessed to contribute to the achievement of the MDGs [Millennium Development Goals]; namely: promotion of gender equality and *empowerment* of *women*... To increase *empowerment* opportunities through ICTs; and the participation of youths and *women* in national development (Government of the Republic of Zambia 2006, 1, 47).[2]

In the case of Uganda, empowerment is used in the Uganda National ICT Policy (2003) to indicate that ICTs enable the:

> ...creation of opportunities and *empowerment* by provision of access to local and global markets and promotion of rural development (Government of Uganda 2003, 8).

At a global level, the term has been used in the World Summit on the Information Society (WSIS) Declaration of Principles (2003), particularly in relation to women and the poor.

> We affirm that development of ICTs provides enormous opportunities for *women*, who should be an integral part of, and key actors in, the Information Society. We are committed to ensuring that the Information Society enables *women's empowerment* and their full participation on the basis of equality in all spheres of society and in all decision-making processes. To this end, we should mainstream a gender equality perspective and use ICTs as a tool to that end.
>
> We are resolute to *empower* the poor, particularly those living in remote, rural and marginalized urban areas, to access information and to use ICTs as a tool to support their efforts to lift themselves out of poverty (WSIS Declaration of Principles 2003).

What does the term "empowerment" imply, and can women, particularly those in rural areas, really achieve empowerment merely through access to and use of ICTs like the Internet and cell phones? Although empowerment is a term used in different and varied contexts (Oxaal and

Baden 1997; Longwe 1997; Sharma 2001; Kabeer 1999 and 2001; Malhotra, Schuler, and Boender 2002; Huyer 2006), several authors look at it from the perspective of women. For instance, Sharma (2001, 82) states that:

> For individuals and groups where gender determines their access to resources and power, their empowerment begins when they not only recognize the systematic forces that oppress them, but act to change existing power relations. Empowerment, therefore is a process aimed at changing the nature and direction of systematic forces which marginalize women and other disadvantaged sections in a given context. The goals of women's empowerment are to challenge patriarchal ideology, to transform the structures and institutions that reinforce and perpetuate gender discrimination and social inequality and to enable poor women to gain access to, and control of, both material and informational resources.

Although Sharma's definition that includes "individual self-assertion, collective resistance, protest and mobilization" is apt, the oppressed are in some cases either too powerless to change power relations or feel that their position is better served without having to act to improve it. While concurring with Sharma, Longwe (1997) emphasizes a strategy of collective action, even with men pursuing women's causes. Empowerment takes into account "the interests of the disenfranchised" (Kabeer 1999, 223), enabling them "to make strategic life choices where previously there were none" (Kabeer 2001, 18). Similarly, Huyer and Sikoska (2003) foregrounds choices and decision-making. But how is empowerment conceptualized?

Conceptualizing Empowerment from the Perspective of Power

Scholars such as Oxaal and Baden (1997) reveal that empowerment stems from the term power. However, this is not necessarily women gaining power over men. Power can be looked at as *power over, power to, power with or power from within* (Moser 1993; Kabeer 1994; Oxaal and Baden 1997; Rowlands 1997; Huyer and Tatjana 2003). *Power over* involves a relationship of domination and subordination, which may include intimidation and violence. Empowering one group implies the other losing power, but Kelly (1991) challenges this aspect, preferring shared power. As opposed to *power over*, empowerment should be *power to*. To Oxaal and Baden (1997), empowerment should involve going a step further to include the aspect of *power with*—organizing with a common purpose to achieve a collective goal. But for this collective empowerment (Rowlands 1997) to be possible, individual members of a group have to be empowered, to have *power within*—recognizing their abilities and being self-confident "to influence and change this situation" (Oxaal and Baden 1997, 1). Ideas of power should therefore benefit both women and men.

Using a feminist standpoint in considering ICTs as tools for empowerment, it can be argued that only by sharing ICT benefits can a more equitable society be achieved. In this regard, the women's movement in Africa,

> ...at its deepest...is not an effort to play "catch up" with the competitive, aggressive "dog-eat-dog" spirit of the dominant system, [but]...rather, an attempt to convert men and the system to the sense of responsibility,...and rejection of hierarchy...(Oxaal and Baden 1997, 2).

Like "empowerment," power is understood differently and has been a subject of debate within social sciences. This dispute is evident in the ways Foucault (1982), Hartsock (1985), and Lukes (1974) discuss power. For Lukes (1974), power is about decision-making and preventing conflict, while Hartsock (1985) looks at power as raising people's morale. For Foucault, power is relational and exists only in its exercise or in a form that if one possessed it the other person would have less of it (Foucault 1982). Feminist critics such as Yuval-Davies (1994) and Todaro (1994) challenge these views and go beyond the tangible expression of power to understanding how gender disparities restrict women's exercise of power, which takes on different forms—inevitably impacting on the different perceptions of empowerment.

This view, however, embodies challenges to patriarchy because it challenges existing social and political structures (Huyer and Tatjana 2003). The idea of power is clearly at the core of the concept of empowerment. So while the concept of empowerment is embraced by some as a key to gender equality (UNDP 1993; World bank 1994; Maguire 2001), others would rather avoid the term because of its association to a shift in power. "Empowerment" requires that power be changed and expanded so that those normally left out can share in it (Batliwala 1997). Empowerment should therefore be about redistribution of power, whereby those who initially were powerless, in this case women, are allowed to share in the power through their increased participation in decision-making processes and control of key resources, thus creating a situation of power to, power within, and power with, as opposed to power over (Moser 1985).

Feminist understanding of empowerment also includes the idea of "the personal as political," therefore calling for a look at empowerment not only at the public level, in this case "relational" or "collective" level, but also at the "personal" level (Rowlands 1997, 14). In the *personal* aspect, women develop "a sense of self and individual confidence and capacity" and undo the "effects of internalized oppression" (Rowlands 1997, 14). The *relational* aspect would involve "developing the ability to negotiate and influence the nature of a relationship and decisions made within it," while the *collective* aspect would involve women working "together to achieve a more extensive impact than each would have done alone" (Rowlands 1997, 15). Empowerment could also occur at an economic, political, or social level. According to Clement (1994), economic empowerment has to do with access to and control of resources. Social empowerment, on the other hand, has to do with challenging social and cultural structures. He argues political empowerment does not come easy because it involves the rights and abilities of people to participate as equals in decision-making processes, but it could be attained through individual and collective action.

Another view to the empowerment approach, put forward by the Development Alternatives with Women for a New Era (DAWN),[3] is that addressing women's strategic needs and practical gender needs is important for breaking down structures of subordination, but women have to struggle collectively to overcome inequalities (Longwe 1997). "Practical gender needs" are immediate necessities such as food, clothing, shelter, and healthcare, while "strategic gender needs" are necessities that exist as a result of disparities between genders. This is in reference to women's subordinate social position in relation to division of labor, power, and control over resources (Moser 1993).

Malhotra, Schuler, and Boender (2002) conclude that although there may be different concepts of empowerment, key to these are four principles for women—options, choice, control, and power over their lives. That is, the ability and the potential to make choices and take control over their lives, the power to make decisions and options that pertain to the betterment of their lives at political, social, or economic levels.

In the context of ICTs, it would appear that in order for ICTs to empower women, they would need to effect Malhotra, Schuler, and Boender's (2002) four concepts. In this regard, Huyer (2006, 27) states that

> for women to be active agents in their empowerment in the knowledge society, as a result of independent and informed choices, it is required that women become creators and developers of the technology and the content it carries. [There is need therefore] to promote the ability to access, use, create, and distribute knowledge.

This promotion can be seen in a number of community gender-based ICT projects in both Uganda and Zambia, which particularly target women. Also, they can be seen in the type of ICTs they choose to engage with the most when it comes to choosing between the Internet and cell phones.

In addition, in considering Clement's, DAWN's, and Moser's concepts, rather than mere rhetoric, policymakers would do well to mainstream gender at all ICT levels. Although governments in Africa point out the importance of gender balance in ICT use and access, their private-sector "partners" have no such concerns. For instance, the National ICT Policy of Zambia lays down no stipulation that private service providers must engage with gender concerns, for example, as part of a tendering process. This lack of gender compliance is extremely problematic; all partners/stakeholders need to engage with and show commitment to gender equality.

To realize Clement's take on empowerment, policymakers and shapers—including the private sector and donor agencies—need to display some form of gender concerns, including an awareness of women's practical needs as suggested by DAWN. This may include considering women's economic capabilities, their education levels, jobs, and so on in order to arrive at gender equality. Such considerations make other stakeholders, other than governments, realize that ICTs cannot be implemented as if they were neutral, because there are control aspects over resources, including aspects of power and the division of labor, which need to be recognized before ICTs can play a very meaningful role in the lives of users, especially those of women.

Measuring Women's Empowerment

Having discussed the above, there is always a need to try and measure, evaluate, or show targets or indicators of whether, and how far, a situation has improved or change taken place. Therefore, measuring women's empowerment is no exception in this respect. However, empowerment, as scholars such as Alsop, Bertelsen, and Holland (2006), as well as Oxaal and Baden (1997), have found, is a difficult concept to measure, not only because it is difficult to quantify but also due to the difficult-to-predict dynamic processes and changes. Oxaal and Baden (1997) state that although there are ways of developing empowerment indicators, due to the multifaceted nature of empowerment, such indicators will not be a whole reflection of empowerment because the concept itself is not easy to quantify. However, the fact that measuring empowerment might not give a complete picture has not stopped scholars or agencies interested in the concept from coming up with measures. For example, Malhotra, Schuler, and Boender (2002) give different examples of frameworks for measuring women's empowerment by different scholars. The multifaceted nature of women's empowerment also means that it is not one-dimensional, such that although a woman might be empowered politically, it will not always follow that she will be empowered financially, and vice-versa. Also, the fact that some women might have the power to access the Internet will not necessarily guarantee them the power to download any type of information, because they might not have the financial means to do so. Therefore, being able to access and use ICTs does not mean unlimited access to all information.

Back to the ideas of "change" and the power to make "choices," Kabeer (2005, 18) identifies three elements for making choices—resources, agency, and achievements. Resources form the conditions under which choices are made, agency is the core element through which choices are made, and achievements are the outcomes of choices made (Batliwala 1991; Longwe 1991; Chen 1992; Sen 1999; Kabeer 2001; Malhotra et al. 2002).

These same ideas are evident in Longwe's (1991) framework, in which she argues that unless the five gender equality levels of welfare, access, conscientization, participation (mobilization), and control are addressed, women cannot be said to be truly empowered. "Welfare" addresses the lowest level at which a development intervention may hope to close a gender gap, while "control" is a level at which there is a balance of power between women and men and neither has dominance. "Access" considers equality of access to resources, and "conscientization" is a level of awareness-raising (about inherent structural and institutional discrimination and the role that women themselves often play in reinforcing the system that restricts their growth). "Participation" is concerned with the extent to which women have been able to take part in decision-making processes alongside men.

Key Results from the Case Studies

Access and Relevance

Socially, communication is easier not only for the club members but also for their communities as a whole due to cell phone use. When it came to considering the access and use of either the Internet or cell phones, it was found that the cell phone was what the clubs and the local communities used the most. One reason for this was the lack of knowledge and understanding as well as reason to use the Internet. Another was due to Internet's unreliability owing to poor reliability of the dial-up connection. The following is an extract of an interview conducted in Zambia:

> About the internet I have heard and also kind of understand how it works but...a lot of people don't have the internet, they use mobile phones at least....in many sections where we live,...there is no one with internet. It's only those who get a lot of money who know about

> the internet. A lot of people know about the mobile phone but not about the internet.

Explanation for this choice is that users found the cell phone more relevant to their daily lives, so they were inclined to use it more and put more money in it. Therefore, the issue with the Internet is perhaps not so much about affordability, but how to use it and what to use it for, and whether its use is necessary. If this could be made much clearer, perhaps via government awareness campaigns, its usefulness might manifest much more easily, especially among poor and marginalized communities that mostly include women.

Additionally, Internet usage within both the Kalomo Bwacha Women's ICT Club and CEEWA-U projects was rarely enquired into by clients, yet it was the Internet that members of the clubs and their benefactors felt would be a cornerstone in achieving social and economic development as well as political empowerment. The Internet did not turn out to be a relevant ICT for the particular communities. Part of the problem for this lay in the lack of consultation between the donors of the ICTs and the recipients. It was assumed that the Internet would be an ideal ICT for empowerment; there was no consultation on what ICT would be relevant for the club and subsequently for the community, and how this would result in economic empowerment. Additionally, the cell phone played a crucial role in social empowerment, given its use to communicate with relations living away from them. Such communication was mainly to inform them about illness or funerals in the family, as well as to ask for assistance for such needs as fertilizers.

ICTs for Women's Political Empowerment

A few of the women who participated in the projects, particularly in CEEWA-U, moved on to participate in the local politics of their sub counties and are now councilors. They say their confidence level has gone up and they can now participate in meetings and demand for accountability, something they say they were unable to do before participating in the ICT project:

> I used to be timid but with CEEWA-U. I have learned to be confident in public; now I participate in meetings, I contribute ideas and ensure the budget for women's issues is increased.

There was, however, very little evidence of the use of ICTs for political empowerment. This is perhaps due to the fact that ICTs have not been effectively used within political realms. Therefore, it would be difficult to expect grassroots communities to use ICTs where there are no visible

ICT gateways to political platforms. For instance, there seem to be no adequate e-governance strategies in place. Although these may begin to exist within government institutions, they are far from being filtered into local communities, let alone into rural areas. However, the Ugandan and Zambian governments could provide forums via the Internet to encourage those in rural areas, such as the club members, to participate in policy consultations. The governments could also perhaps install dedicated cell phone lines, free of charge, for the public to communicate with local councilors or other government officers. The cell phones could also be used for conducting opinion polls. Such measures would encourage a sense of involvement in policy-making and community consultation, particularly for women.

Policymakers make decisions (top-down) that have an impact on women (and men), without necessarily having any effective input from those whom such decisions affect. Due to this, women cannot effectively lobby policymakers through the use of the Internet, or indeed the cell phone. This is difficult particularly for women in rural areas due to connectivity and cost barriers. Additionally, even if one has a handset, it would be difficult to expect the women to use their cell phones to lobby policymakers, as air time is very expensive. Although women do network using the cell phone, this networking is undertaken mainly for their own and their community's immediate social and economic needs—mainly within clubs. Most important, though, without awareness, women—especially those in rural areas—will not realize the importance of the Internet or indeed cell phones as tools for political empowerment.

In terms of power, participation, and decision-making, ICTs did not bring about much change for women at the household, community, and institutional level. The financial and technical support for the beneficiaries influenced the programs implemented, and according to the women, most of what they actually needed, such as a photocopier in the Zambian case, was not among the benefits provided. In addition, the gender relations within the household and institutional level did not change much. Despite access to some ICTs, women still experienced discrimination based on their gender. Some of the women pointed out that that they tried to get loans from microfinance institutions but they always required their husbands to be involved.

In some homes the usage of cell phones sometimes acted as tools to reinforce gender power differentials. For instance, some of the women indicated that as a result of receiving SMSs or calls from people their spouses suspected to be their lovers, they had experienced both verbal and physical abuse, often from their male spouses. Police reports based on actual experience also highlight the fact that the use of cell phones, especially SMS, has contributed to increased domestic violence (Police Superintendent in Charge of Family Affairs, Uganda 2004). If women made use of all the facilities within a phone, say the camera where available, it could help get evidence of violence committed against them, which can help in the arbitration of domestic violence cases.

Co-ownership of Cell Phones as an Indicator of Individual and Collective Empowerment

Aspects of agency at the individual and collective level were evident in the way the women developed strategies to deal with their oppressive situation. Women now co-own phones. They put funds together and buy phones. In Uganda, the co-owned phone would be left in the custody of those who are either single or have more open-minded spouses. Some women, such as Nambi in Buwama, possess phones but their husbands are unaware of it:

> I fear that if he does, he will start monitoring my movements; he could even take the mobile [phone] away from me

In much similar fashion, and rather defiantly, a female interviewee from Zambia who experienced abuse by her husband because of a cell phone, and was forced to sell it, said she had decided to buy another phone because she realized its importance.

Partnership as a mechanism of survival and sustainability of achievements has helped women deal with oppressive situations and, to a large extent, limit men's control over their resources. In addition, co-ownership of phones has also enabled women to deal with cost issues. The cell phone has enabled women to get in touch with suppliers or middle men to arrange for the pickup of produce from their farms. According to Awoko, a fish trader in Buwama, Uganda:

> The mobile phone has been revolutionary. Today I can just call my customers all the way from Kampala to come and collect fish. Before, I used to have to incur an extra cost traveling to Kampala to look for markets or take the fish personally.

For other women the phone has raised their level of self confidence because it has elevated them socially:

> Before the telecenter, some of us had never used a telephone in our lives. We thought things like the mobile [phone] was for only the highly educated or those in Kampala with a lot of money. But CEEWA showed us that we could also use the mobile for many purposes…, now I am confident, I have a mobile and I can use it from my home without walking to the centre.

On the whole, the women are happy with mobile phones and they find it easy to use it, because with the phone one can use whichever language one wants, as opposed to the Internet, which requires knowledge of specific languages, especially English.

Clearly, cell phone usage has enabled women to achieve a number of things, some of which could be indicators of empowerment while others are not. The Internet, however, did not offer much for them, and since women had no access to the Internet outside the telecenter, it was not possible to see how else they utilized these facilities. Just a few women had hopes of being able to use the Internet to do business. An example is Awoko of Uganda:

> From my savings I hope to expand my businesses. If God helps and electricity is extended to remote place like the front where we fish monger stay then perhaps telecenter could be set up there too and I can access computers and the internet to do lots of things for my business, like ordering for goods or even accessing international markets.

Similarly, Bwacha's Chairperson in Zambia optimistically said this of the Internet:

> The Internet might help in that we would have whatever they [the club members] are making advertised and then if there are some interested individuals then they can make some orders. Otherwise they just make but where to sell? . . . that's why we thought of including the internet.

The women also indicated that they had become aware of their individual rights and autonomy to define their personal needs. For example, one CEEWA affiliate said:

> From the CEEWA, I learnt to stand up for my rights, I realized I had been wasting my money supporting my husband's other wives and their children. What I decided to do instead was to use that money to take my child to a good boarding school.

It is, however, worth noting that the projects were not just about using ICTs, because women also received training in business management, public speaking, and were made aware of their rights. Most beneficiaries were already active business women and had also participated in other projects within the community. Their agency and achievements can therefore not be attributed to just the use of ICTs. It is clear that other variables in the women's context also contributed to the changes they were experiencing.

Concluding Remarks

This chapter has looked at the aspect of ICTs in the context of analysis of gender differentials in terms of empowerment and constructivism. It has also considered empowerment, particularly of women, through democratic theory by subsequently giving an analysis of power as it informs empowerment. The cell phone has been identified as the technology most used, especially for those in rural areas who are very unlikely to use the Internet for a variety of reasons, which include infrastructure, cost, skills, knowledge, and awareness. After analyzing power and empowerment, it is fair to note that power affects women and men differently. In order for women to have power to make strategic life choices, which also includes the choice to constructively be a part of the decision-making process in ICT policymaking and implementation, as well as its use, they not only need to have the agency to do so but also the resources. This is possible through collective action, sharing of power, and having policies that not only target the government but also other stakeholders with whom the government gets into partnership. In addition, rather than the policymakers deciding about whether women should confine their ICT participation to community centers or telecenters, how much funding should be allocated to women's ICT training, how many women should be appointed to Ministerial positions, and so on, empowerment will require a change in the status quo in which power lies with one group over another. As such, it should be *power to* or *power with* others instead of *power over* others. Women's resources in all forms will therefore need to be taken into serious consideration for them to realize any form of empowerment. For as Sharma (2001) ably puts it:

> Women's empowerment should lead to the liberation of men from false value systems and ideologies of oppression. It should lead to a situation where each one can become a whole being, regardless of gender, and use their fullest potential to construct a more human society for all (Sharma 2001, 87).

For without realizing the above there can be no real development, particularly as more than half the population remains disempowered. To this end, when women's potential is recognized and utilized equally, politically, socially, and economically, development for everyone can ensue fairly.

Notes

1. Most women are confined to the informal economy.
2. Italics are our emphases in this and the following quotes.

3. The DAWN concept arose out of an NGO forum in 1985 to offer Southern women an alternative development model of social advancement that is equitable, participatory, holistic, and sustainable and respond to people's needs. Its philosophy reflects the dissatisfaction of many developing countries at the time, with unfavorable terms of trade, protectionism, and the conditionalities of the World Bank's Structural Adjustment Programs.

References

Burgess, A. 2004. *Cellular Phones, Public Fears, and a Culture of Precautions*. Cambridge: Cambridge University Press.

Butler, R. 2007. Cell Phones May Help "Save" Africa. http://news.mongabay.com/2005/0712-rhett_butler.html (accessed February 11, 2007).

Cairncross, F. 1998. *The Dearth of Distance: How the Communications Revolution will Change our Lives*. London: Orion Business Books.

Castells, M. 1997. *The Rise of the Network Society*. Cambridge, MA; Oxford, UK: Blackwell Publishers.

CAZ. 2004. *2004 End of Year Economic Requirements*. Lusaka: Communications Authority of Zambia (CAZ).

CEEWA-U. 2005. *Women and Entrepreneurship Development Program-ICT Project: Annual Narrative Report*. Kampala: Council for Economic Empowerment for Women of Africa-Uganda.

Clement, A. 1994. "Computing at Work: Empowering Action by 'Low Level Users.'" *Communication of the ACM* 37: 53–63.

Cockburn, C. and S. Ormrod. 1993. *Gender and Technology in the Making*. London: Sage Publications.

CSO. 2004. *Living Conditions Monitoring Survey Report 2002–2003*. Lusaka: Living Conditions Monitory Branch, Central Statistical Office (CSO), Government of the Republic of Zambia.

Faulkner, W. 2000. "The Technology Question in Feminism: A View from Feminist Technology Studies." *Women's Studies International Forum* 24: 79–95.

Government of the Republic of Zambia. 2006. *National Information and Communication Technology Policy*. Lusaka: Ministry of Communication and Transport.

Hartsock, N. 1985. *Money, Sex and Power: Towards a Feminist Historical Materialism*. Boston: Northeastern University Press.

Heap, N., R. Thomas, G. Einon, R. Mason, and H. Mackay. 1995. *Information Technology and Society*. London: Sage.

Horst, H. A. and D. Miller. 2006. *The Cell Phone: An Anthropology of Communication*. Oxford and New York: BERG Publishers.

Huyer, S. 2006. "Understanding Gender Equality and Women's Empowerment in the Knowledge Society." In *Cinderella or Cyberella? Empowering*

Women in the Knowledge Society, ed. N. J. Hafkin and S. Huyer, 15–49. Bloomfield: Kumarian Press.

Huyer, S. and T. Sikoska. 2003. Overcoming the Gender Digital Divide: Understanding ICTs and their Potential for the Empowerment of Women. *INSTRAW Research Paper Series* 1. http://www.onlinewomeninpolitics.org/beijing12/2003_gender_ict.pdf (accessed December 27, 2008).

ITU. 2008. Africa, ICT Indicators, 2007. Updated April 24, 2008. http://www.itu.int/ITU-D/ict/statistics/at_glance/af_ictindicators_2007.html (accessed September 16, 2008).

Ling, R. 2004. *The Mobile Connection: The Cell Phone's Impact on Society.* Boston: Morgan Kaufmann.

Malhotra, S. and C. Boender. 2002. *Measuring Women's Empowerment as a Variable in International Development.* World Bank Workshop on Poverty and Gender: New Perspectives.

Moser, C. 1993. *Gender planning and development: Theory, practice and training.* New York: Routledge.

Murelli, E. and R. W' O Okot-Uma. 2002. *Breaking the Digital Divide: Implications for Developing Countries.* London: Commonwealth Secretariat and SFI Publishing.

Oxaal, Z. and S. Baden. 1997. *Gender and Empowerment: Definitions, Approaches and Implications for Policy.* London: BRIDGE (development-gender) Institute of Development Studies, University of Sussex.

Rowlands, J. 1997. *Questioning Empowerment: Working with Women in Honduras.* Oxford: Oxfam Publications.

Roy, S. 2005. *Globalization, ICT and Developing Countries: Challenges in the Information Age.* New Delhi, Thousand Oaks, and London: Sage.

Sharma, U. 2001. "Men are Real, Women are 'Made Up': Beauty Therapy and the Construction of Femininity." *Sociology Review* 49: 100–116.

UCC. 2006. Communications Sector Comparative Figures for the Period December 1996 to March 2006. Kampala: Uganda Communication Commission (UCC).

United Nation's WSIS Declaration of Principles. Building the Information Society: a Global Challenge in the new Millennium. http://www.itu.int/wsis/docs/geneva/official/dop.html (accessed June 26, 2008).

Wajcman, J. 1991. *Feminism Confronts Technology.* Cambridge: Polity Press.

Yuval-Davies, N. 1994. "Women, Ethnicity and Empowerment." *Feminism and Psychology* 4: 179–197.

CHAPTER 8

ICTs, Youths, and the Politics of Participation in Rural Uganda

Carol Azungi Dralega

Introduction and Research Problem

This chapter[1] explores the participatory ethos and rhetoric around information and communication technology (ICT) access and use among rural youths in Uganda for democracy and development. The main discussions of this study derive from the perception that democracy constitutes an essential part of development and the notion that young people are simply "future potential" and receptacles for learning need to be challenged. Instead, youths should be perceived and treated as active, contemporary, and indiscriminate actors in their own right.

The study's focus on rural youth is not just because they form the majority of the youth population in Uganda (Uganda Bureau of Statistics 2007), but also because they belong to communities that are in the main both geographically and socially peripheral. They live in communities that also happen to be conservative in their attitude toward the introduction and usage of new technologies and are also impoverished, and consequently have limited access to ICT resources.

This investigation is based on the (Nakaseke) Community Multimedia Centre (CMC) model for universal access that combines telecenter facilities (such as telephone, fax, library, Internet, and photocopiers) with community radio. The establishment of the Nakaseke CMC (about 50 kilometers from the capital city, Kampala) was therefore seen as a communicative platform where the youths of the community could actively contribute and participate in mainstream discourses, which generally overlook minority concerns.

Using "participatory approaches" to development communication as the overarching analytical frames, the goal is thus to reexamine the project's initial aims to revitalize rural engagement in public discourses by comparing it with the reality on the ground. In doing this, the study explores how youth civic engagement within a specific political context reacts with the Community Multimedia Centre principles, particularly access, ownership, and sustainability. The chapter highlights the specific linkages between media control, public communication, national electoral politics, and market mechanisms that are conjectured to weaken political debate at the expense of democratic development. In particular, the study points to a shift from active politics to apolitical engagement and argues that the subsequent rhetoric surrounding this shift are particularly indicative of antecedent contextual influences specific to remote and underdeveloped rural communities.

The study sets out to answer the following key research questions:

- How has the Nakaseke CMC (model) approached the questions of ICT access, participation, and inclusion of youths in ICT and civic engagement?
- What, consequently, are the outcomes and implications of these processes?

Research Context

Youth and ICT Discourses in Africa

It has often been argued that the future of any society depends on the successful provision of well-marked pathways where young people prepare themselves to become contributing adults in their communities (Brown, Larson, and Saraswathi 2002). African demographics indicate that the youth represent 60 percent of the continent's populations—a statistic, almost opposite of those for other continents (ADF-V 2006). The African Development Forum (2006) projects that by 2015 the youth proportion of the total African population will exceed 75 percent, yet these demographic facts are not adequately reflected in Africa's democratic and development processes, including access and ownership of information resources like ICTs.

Statistics of ICT access in Africa are a manifestation of the global digital divide, with repercussions on the national and democratic divides that directly affect ICT access and use among rural youths in Africa (Norris 2001). In terms of the fixed telephone market, with 28.5 million lines in 2006, Africa had a share of just 2 percent of the global fixed telephone lines. Almost 80 percent of all fixed lines in Africa were concentrated in just six of Africa's fifty-four economies (Algeria, Egypt, Nigeria, South

Africa, and Tunisia), and were "almost exclusively located within the cities and randomly in rural areas" (ITU 2007, 8). Africa's penetration was about a tenth of America's and Europe's and about a sixth of the world average in 2006. Regarding cell phones, Africa recorded significant growth—with higher cellular subscription than the fixed line subscription and with more income generated from cell phones than from fixed telephones (ITU 2007). However, Africa's cellular subscription was still very low compared to the rest of the world, where markets have reached saturation. In terms of Internet access, less than 3 percent (10.7 million) of the world's Internet subscribers were located in Africa in 2006 (ITU 2007).

However, despite these disparities in global access, ICTs have become a significant factor for development in many African countries, and have a profound impact on the political, economic, and social sectors, presenting both opportunities and challenges in terms of the social development and the inclusion of youth. In fact, as Mihyo and Sesan (2004, 132) infer, ICTs are an area where young African people have an edge over their adult counterparts, an edge that over the last thirty years threatens to shatter the myth of adult supremacy and monopoly over the world of knowledge.

Kinunda (1999) argues that while the process started in the conventional domains of learning with the introduction of youth-to-youth learning processes relating to community development, child hygiene, and child safety, ICTs have tipped the scale. ICTs have put youth in the centre of the training processes, as adults have been slower in acquiring the new techniques. The youth presence in Internet cafés and community telecenters in Africa is overwhelmingly dominant. This includes their roles giving instructions to their peers and to adults, running cell phone shops, providing IT support services, including repairs, and also as IT teachers in schools, where most of the IT teachers are young (Soltane et al. 2004).

A clear manifestation of this trend can be found in the results of a survey carried out by Mwesige, who explains the dominance of youth in the IT field in Uganda: "A typical Internet café user in Uganda is a 25-year-old single male with no children, who has completed high school at the very minimum. Eighty percent of café users were under 30, and less than 4 percent were over 40 . . ." (Mwesige 2004: 93–94). The current research also indicates that youths are the dominant ICT users at the Nakaseke CMC, whose manager acknowledges as much:

> Youths are the biggest users of the CMC and therefore represent the biggest source of revenue for us as compared to their adult counterparts . . .[2]

Although the effectiveness in terms of the developmental and democratic uses of ICTs by the youths of Africa is debatable, Mihyo and Sesan

(2004, 132) acknowledge that young people in Africa are often the leading innovators in the use and spread of ICTs because they adapt quickly and are generally very eager for the great quantities of information available locally and globally. Indeed the African Development Forum (ADF-V 2006) notes several potential economic advantages of a youthful population, and highlights their creative value. For this reason, the youths have a legitimate claim to be partners in the developmental and democratic processes.

Until recently, youths in Africa have generally been marginalized in national economic, social, and political structures. However, in 2000, at the UN Millennium Summit, 147 of the world's political leaders, including many African Heads of States, signed a declaration that ushered in the UN Millennium Development Goals (MDGs),[3] and all of them related to youth, although only one, the eighth, made specific mention. The youth are vital due to their numbers and their historical contribution to Africa's political history (Soltane et al. 2004), despite confusion and conflicting perceptions—some seeing the youth as a threat to the established order while others focusing on young people as the hope for the future (Bell 2005). To some extent, discourses on African youth and ICT use indicate an increased emphasis on using ICTs in the context of global youth priorities, such as access to education, employment, and poverty eradication, going by the examples of Kenya,[4] Paradigm Initiative Nigeria,[5] and the African Youth Initiative[6] (Mihyo and Sesan 2004, 132–136) as a very small representation of youth ICT initiatives.

The youth of Africa are increasingly making use of possibilities provided by new technologies to access entertainment and news sites, and as personal meeting spaces through chat programs. Despite these varied activities, a couple of questions about participation and benefits still remain unanswered. The first question is concerned with youth participation in a number of political civic activities and the second queries whether ICTs have the capacity and are capable of adequately empowering young people and improving their lives.

ICTs, Youths, and the Rural Divide

Youths in grassroots communities form a significant group of interest in the agency of ICT discourses in Africa in general and in this study in particular. Of the limited academic research concerning rural youth, one by Mwesige (2004) offers some insights into the rural-urban ICT divide. His findings indicate that 90 percent of Internet users are urban-based, educated, young, and male—highlighting the dilemma faced by rural youth in terms of ICT access. He argues that "while initiatives such as cyber cafés have brought the internet and ICT closer to more people in developing countries,...that these initiatives, especially when they are commercially based, may only be increasing the digital divide within poor countries" (Mwesige 2004, 17).

The question of ICT access and civic participation among local community youth is linked to a number of practical problems that are particularly relevant in the poorest areas of the world. These include the national dimension, which comprises the macro-level technological, socioeconomic, and political environment within which the ICTs are accessed; the institutional factor, which helps explain the opportunities offered by the structures and operation of the ICT access institution; and finally the individual dimension, indicative of the micro level of resources and motivation among youths as a determinant of participation. These three dimensions will be explored as a framework for explaining the digital divide and notions of access and participation in Nakaseke.

Contextualization of the Nakaseke CMC

The contextualization of the Nakaseke CMC sets the stage for the analysis of the findings, which derive from the founding ambition of Uganda's first multi-donor funded, Multipurpose Community Telecenter (MCT) project launched in 1999. The main ambition was to demonstrate how ICTs could enable grassroots and underserved communities to solve developmental problems and build effective approaches to access and disseminate knowledge (Acacia 1998). Two dominant visions underscored this ambition, the first was to provide rural and remote communities with *access* to ICTs and communicative spaces hitherto absent, and the second was to revitalize the concept of *participation* in democracy and development among peripheral communities, with efforts made to address the different demographics of the community.

The donation of a "suitcase" FM radio transmitter by the Uganda National Commission for UNESCO as part of a Network of Pilot Projects (that is Buwama, Nabweru, Apac, Kagadi, and Kacwekano) converted Nakaseke's MCT into a CMC in 2002. The implementation process was joined by several collaborative institutions, which included the Department for Information in the Office of the President, the Uganda National Council for Science and Technology, and the National Libraries Board. Although Nakaseke operates autonomously, like the other CMCs, inter-CMC synergies are expected to enhance benefits from common strategies for communication, support, cost-effective networking solutions, and shared resources in areas such as ICT expertise, technical support, information sources, local management, training, support agencies, and social marketing.

In terms of its organizational and operating principles, Nakaseke is an independent community radio and telecenter type of CMC, in the sense that the community radio station shares the premises and all the management and other structural arrangements with the telecenter facility. The radio usually broadcasts in FM from 8AM to 6PM daily within a radius of twenty-five kilometers.[7] The centre is staffed mostly by thirty-two volunteers and four permanent staff. It earns some income from announcements, messages, and programs paid for by individuals and organizations. The telecenter has eleven computers for public use, although only five were functional at the time of the research. It charges for Internet access, scanning and photocopying, as well as for training courses. It also offers some services free or at discretionary rates to particular groups within the community, according to community needs and development priorities. The use of radio together with other print and nonprint information was seen to provide a basis for enhancing effectiveness and reach to disadvantaged communities.

A Local Management Committee (LMC) board was established consisting of various group representatives—farmers, women's groups, and local government—in addition to youth representatives selected by the community youth whose main role is to provide feedback to the management on issues of concern to their constituencies. What the youths' demands were and how they have been met will be discussed shortly.

Theoretical Framework

Development communication theory offers an overarching analytical frame for discussing participatory political engagement and participatory development that arise in this study. Consequently, components of empowerment and participatory communication theory are applied as overall guiding principles to the discussion of civic engagement. The empowerment approach is fruitful as it incorporates the Freirian liberation assumptions in the methodology, which are oriented in social action and works toward the achievement of empowerment-related outcomes to which participatory democracy is attributed (Jacobson and Servaes 1999; Melkote and Steeves 2003; Hagen 1992).

According to Jacobson and Servaes, through development-support communication, "participatory communication is seen as a push away from 'uniform, centralized, top-down, expensive, professional and institutionalized media' to more of 'multi-dimensional, horizontality, inclusive (even friendly), deprofessionalized, and diachronic communication exchange' in which the voices of the marginalized in rural and remote areas are heard, whereby the senders and receivers interchange roles" (Jacobson and Servaes 1999, 84).

While participation, as indicated above, is the preferred approach to rural communication for development, this study contends that participation should be looked at both as a democratic and development imperative, as

Bessette points out:

> Promoting participation also depends on making room for democracy and recognizing the right to express divergent opinions. Without democracy and respect for fundamental human rights, and without the freedom of expression, the ability to use communication to foster social change is severely limited (Bessette 2004, 14).

While democracy as a form of government narrowly focuses on the political machinery, democracy as a kind of society, as perceived in this study, looks at the role of citizens as active participants within that society (Hagen 1992, 17; Rozumilowicz 2002, 11). In the same light, participatory democrats look beyond the instrumental conception of people as voters or choosers of their leaders and view them as active participants in the decision-making processes at all levels (Hagen 1992).

Participatory theorists such as Pateman (1970) and Behrouzi (2005) emphasize the need for direct participation by citizens in the regulation of key institutions in society, including the workplace and local community—including rural and grassroots communities. Hagen supports Pateman by saying, for a democratic policy to exist it is thus necessary for participatory society to exist (Hagen 1992, 18). She suggests that such a process should include the mass media, which play a crucial role in society as a source of information, giving people access to receive, produce, and exchange information (Hagen 1992, 21). For democratic communication to occur, different social groups must have access to the means of communication, where they can be actively involved in program planning, production, and decision-making in general (MacBride Report 1980).

The reasons for these discussions on participation and democracy draw from critical works that relate to key concepts such as participation, representation, discourse, and power that are relevant to discussions on ICT, youth, and civic engagement in rural Uganda. This brings to light Habermas' notion of the public sphere, which remains central in discussions of democracy. Allowing free access to information and competition of different perspectives in the "marketplace of ideas" lead to the triumph of the best idea, enabling citizens to participate in the politics and make informed political decisions.

Though Habermas' account of the public sphere has lost much of its significance and has been criticized variously as "patriarchal" and "colonial" (Chibita 2006), the concept still remains central to the formulations of democracy, citizenship, and public opinion. Participation, free access, deliberation, rational discourse, and a platform from which to question the legitimacy of authority are some of the features of the public sphere, which continue to be crucial to any social formation, including the discussion of youth discourses in Nakaseke.

Also, the discussions draw upon critical political economy theories, an offshoot from mainstream political economy, particularly for its central concern for the balance between capitalist enterprise and public intervention and its engagement with basic moral questions of justice, equity, and public good (Murdock and Golding 2005: 61). In the critical political economy, it is argued that people's capacity to participate in public discourses through the media is related to broader changes taking place in the structure and operation of the media and that this capacity is curtailed by not only material barriers but also by symbolic ones. This theoretical proposition (used sparingly) may help explain market-related discourses evident in the Nakaseke case.

Methodology

The analysis of the empirical material is based on qualitative methods, which are explorative and explanatory (Yin 2003). Data sources on Nakaseke are mostly experiential, spoken, and written (Rossman et al., 1995; Sarantakos 1998). These included focus group discussions (Barbour and Kitzinger 1999), mainly targeting users, and structured interviews (Sarantakos 1998) with the CMC manager and the youth representative on the board of the Local Management Committee. Telephone and e-mail follow-ups have been ongoing (Steward and Mann 2000). Also, the Internet has been a key resource in accessing information, especially as regards secondary data—that is national ICT and telecommunication statistics.

Findings and Discussion

Nakaseke CMC maintains a "no-politics" policy among its users. It is argued here that three structuring and restructuring phases have led to the build-up of this policy. Phase one was the optimistic phase surrounding the establishment of the centre in 1999, explained earlier; Phase two was the "political election window" leading to the ban on politics; and Phase three was in the "proliferation of apolitical engagement." What follows is an explication of these phases and a discussion of their implication in the bid to answer the two key research questions stated earlier.

ICTs, Youths, and Political Engagement

Phase two has its roots in the protracted political history of Uganda, which culminated in the 2005 presidential and parliamentary elections in which active manifestations of political engagement in Nakaseke created active political participation and unexpected reactions from the CMC management. In order to understand the context of this political engagement, a brief political flashback is necessary.

After decades of political turmoil since its independence in 1962, the coming to power of Museveni in 1986 had promised a fundamental change with the restoration of personal freedoms and improvements of the socioeconomic conditions of the Ugandan people (Mutibwa 1992). Since the early 1990s, with the support of the World Bank and International Monetary Fund, Museveni vigorously pushed for economic liberalization and privatization. The 1990s thus saw Uganda recording some of the fastest growth rates in Africa (UNDP 1998; Tangri 1999).

However, the economic liberalization of the 1990s was not accompanied by an equally liberal political environment. In 1995, Uganda adopted a new constitution that entrenched a one-party system. After intense internal and external pressure, the ban imposed on political party activity was lifted with the 2005 presidential elections. Yet several challenges to the current regime remain, which has affected health, education, and rural development budgets (Mugaju 1999; Mbaine 2006; Mwesige 2004). Moreover, there is increasing domestic and external concern over corruption in both public and private sectors (Global Corruption Report, 2003). Despite the relative political stability of the past twenty years, political polarization persists on an increasingly shaky political landscape, which resulted in election clashes[8] in Nakaseke.

The presidential elections in 2005 were therefore characterized by two main factions: one was the incumbent president, Museveni, who after ruling for 20 years had manipulated the constitution to allow for a third term in office under the "no change" slogan. His main opponent was Dr. Kiza Besigye, who called for "change" and had attracted a large share of the electorate, including Nakaseke youths, into active revolution. Consequently, local youths engaged in highly politicized public sphere debates in the community centre media (particularly the radio), which then resulted in violent street clashes with serious repercussions for the very survival of the CMC.

The Banning of Politics at the Nakaseke ICT Centre

Following these highly divisive multiparty clashes during the presidential and parliamentary elections in February 2005, in which youths actively engaged in divisive partisan demonstrations and "unruly" behavior, local politicians and CMC management abolished all political activities and communication at the ICT centre. The argument, according to the manager, was that the CMC

> is a self-sustaining project and engaging in party politics was bringing problems as we were branded segregative by some factions who vowed never to use the centre, therefore denying us revenues...[9]

Instead, the Centre management evoked the notion of neutrality. The manager said they turned down any political activities at the CMC "because it is a neutral service provider whose aim is to serve the whole community without favoring any party."

Although liberal democracy requires a neutral media system, the recourse to the notion of neutrality in Nakaseke raises questions especially in regard to the classic liberal logic that forms the foundation to Habermas' notions of the public sphere. The argument is that a space with a fair degree of autonomy from the state and the market is essential to hold the state accountable to the rest of society (Habermas 1986, 3, 90–120, 248). The problem with Nakaseke was that the political space was compromised in the face of discord and disagreement. We therefore see that although the establishment of the CMC had been interpreted as an attempt to entrench the process of participatory democratization, this does not work in reality when political and market forces come into play. The question then is whether the market-driven enterprise creates a fair society and maximizes the potential for self-realization. It is one of the questions I attempt to discuss in the following sections.

From Active Politics to Apolitical Engagement

Following the ban on political engagement at the CMC, the management decided to focus more on the apolitical engagement of the youth. Based on the feedback obtained from the youth representative to the management on issues of concern, the main demands of the youths from the CMC were to organize talk shows on youth-related issues, provide documentation on issues that affect them, such as HIV/AIDS, alcohol, and organize moneymaking activities that youths could engage in to supplement their parent's incomes, and also conduct social events that would unite and consolidate the youths' sense of community. As a result of these feedbacks, the CMC management developed the following programs.

Muvubuka Weyogerere: The Radio Show for Youths

Muvubuka Weyogerere is a program conceived with the aim to give youths a voice and encourage them to speak out. So, once a week, for one hour every Thursday evening, Nakaseke FM broadcasts the youth-specific program, which translates to *"Youths, Speak Out."* In collaboration with the Uganda Red Cross, the Nakaseke management (with the help of volunteers) collect, package, and disseminate information about HIV/AIDS, as was requested by the youths. This information is multisourced from the Internet (through RadioBrowse), the Ministry of Health, the local government, and the library, as well as researched from the community. During this one hour that is dedicated to the gravest issues that concern youths (such as AIDS), live phone calls are encouraged to the producers who often have expert guests in attendance.

In addition to the *Muvubuka Weyogerere* program, thirty minutes are set aside every Sunday for discussions on youth's issues. According to the centre manager, issues discussed include land ownership and documentation, alcohol and drug abuse, money-raising activities, and education matters. The program is broadcast in *Luganda*, the local language.

From these two CMC programs, several principles of participatory communication stated earlier are discernible. For instance, *Muvubuka Weyogerere* valorizes participatory-grassroots communication, which involves youths in an interactive or two-way communication flow, allows them to identify and voice their civic rights and problems, helps them understand the causes of their problems, encourages them to propose solutions themselves, and then facilitates or encourages them to organize themselves to take appropriate action. The challenge with this approach, according to the manager, was the limited content.

Live Broadcast of Public Events

The geographical isolation of Nakaseke dictates that social life, especially for youths, revolves around church communities, local clubs, and village events. As a consequence of the youths' feedback, the CMC management organizes football galas, usually sponsored by the NGOs dealing with local community issues relating to youths, as well as events sponsored by several radio stations, notably the Kampala-based Central Broadcasting Service (CBS). These live events help raise awareness about activities at the CMC and propagate valuable community-related information around the event.

Drama Clubs as Moneymaking Ventures

In response to the demand that the centre assist the youths in generating an income, the Nakaseke CMC started a drama club with a focus on "out of school" youth with talent but no financial support. The club puts on shows, and the revenue raised from the ticket sales benefits both the actors and the Centre. In addition, the plays dramatize some of the pertinent issues affecting youths mentioned above. Sometimes these shows are broadcast on the community radio.

These approaches of access, participation, and inclusion of youths at Nakaseke are in accord with development communication propositions, which together provide an assumptive basis for participatory strategies. Although several scholars observe that there are different levels of "community participation" (Mato 1999, 59; Jacobson and Kolluri 1999, 268–269; Dervin and Huesca 1999, 176; Jacobson and Servaes 1999, 84), the case of the Nakaseke youths underscores Dervin and Huesca's (1999, 178) argument that in communitarian participatory communication practice "the bulk of recommendations for communication applications take the form of rather modest, specific actions such as: linking up with organizations and civic groups; scheduling a high percentage of programming in indigenous languages; recruiting practitioners from surrounding communities; visiting popular gathering spots when constructing socio dramas and news; incorporating a variety of formats—humor, debate, games, and so on—into programming," which are concepts that have been relatively well adopted by the Nakaseke youth.

Free Peer-to-Peer ICT Training

In addition, through the initiatives discussed above, cheaper ICT training is made possible in which financial concessions are made to accommodate youths with "shallow pockets." Peer-to-peer ICT propagation is encouraged (as lack of ICT skills was one of the problems affecting youths). Peer training engages more youths by including them in ICT discourses, which can be seen as democratic and innovative. This horizontal move signals a positive departure from earlier, overly top-down and prescriptive approaches nowadays shunned in development communication (Melkote 2001, 339).

Diaz-Bordenave (1980) warns that the problem with these new approaches is that the expected participation was often directed by the sources and change agents and that people were induced to participate in self-help activities whose solutions to local problems were already selected by the external development agencies. This may be true to a small extent as some of the information, especially on HIV/AIDS, and drug and alcohol abuse, has been nationally researched and packaged by the change agents: the Ministry of Health, the Red Cross, UNICEF, and other such institutions. But since the community of Nakaseke does not have the capacity to research these issues, the centre understandably obtains information on them from other institutions.

Recourse to the empowerment discourses recommended by proponents such as Ascroft and Masilela (1994), Jayaweera (1987), and Melkote and Steeves (2003), as well as the systemic participatory methodology and the implementation of horizontal processes in Nakaseke, in which youths are directly associated with the communication process where management takes their knowledge and their viewpoints into consideration in the communication process, would constitute major elements of empowerment.

Melkote and Steeves (2003) contend that the key goal to development must be empowerment, whether at the individual, community, or national level. According to them, empowerment is multifaceted in that creative survival requires both material and nonmaterial resources, resources that vary by context. The Nakaseke case conceptualizes development communication in facilitating empowerment outcomes, and as there is no single recipe for facilitating empowerment, these models and experiences, although nonpolitical, offer useful lessons and insights into democratic civic engagement in a (rural) developing country context.

Consequences of Apolitical Engagement to Democracy

Nakaseke has made attempts at civic nonpolitical engagement, which means the issue in Nakaseke is not how to engage youths but rather how to engage them in nonpolitical issues that affect them in order not to upset the political and economic equilibrium of the CMC and the community.

Nakaseke unravels the fundamental conflicts arising between reality (at the time of this research in 2005–2006) and utopia ("idealistic" visions at the inauguration of the telecenter in 1998) in political engagement regardless of the contexts of geopolitical and socioeconomic situations. As mentioned earlier, the initiative at Nakaseke was originally aimed at energizing civic participation among youths in the community by providing them with access to information, knowledge, and horizontal communication possibilities. However, the CMC management, challenged by the outcomes of political engagement, was unwilling to open any political spaces and consequently the center management ended up restricting the foundations of their visions of access, freedom of speech, and political engagement. The rhetoric of local public sphere engagement in Nakaseke negates participatory communication notion.

Whereas the introduction of CMC in Nakaseke had been interpreted as an attempt to entrench the process of democratization, the move to decentralize the media does not necessarily lead to participation in democratic political engagement. Actually, the case of Nakaseke supports elitist theories that argue mass participation would upset the core stability of the democratic system. This seems to conflict with participatory democratic theorists who mount a strong critique of elitist theories (Hagen 1992, 18). The answer to this dilemma may lie within the critical political economy perspective to participation.

As stated earlier, critical political economy holds that people's capacity to participate in public discourses through the media is related to broader changes taking place in the structure and operation of the media (Murdock and Golding 1997; Murdock 2000). This is true in Nakaseke as the voices for neutrality are louder when notions of free-market principles are included in the debate. These voices resonated in Nakaseke as the manager's call for neutrality derived from the fact that the center was an open-market enterprise, and siding with one group would lead to the loss of customers from the opposing side. Neutrality in Nakaseke was vital since the center was faced with the harsh realities of sustainability, which threatened its survival. For the case of Nakaseke then, resorting to apolitical engagement was a means to include youths into ICT discourses without upsetting the survival of the centre, even if that meant limiting youths' rights to political engagement.

Nakaseke and the Question of the Rural Digital Divide

Discourses on the digital divide have produced the participation hypothesis, according to which ICTs influence the involvement of young people, as individual or micro level of resources and motivation among youths is a key determinant of participation (Norris 2001: 12). Norris refers to this as the ***democratic divide*** between those who do and those who do not use any of the multiple (political) resources available for civic engagement.

According to the CMC manager in Nakaseke, although youths comprised the biggest users of the ICTs, not all were active in the selected civic activities. Many of the youths who attended or called the radio program were regular participants. Moreover, most of the young Internet users preferred individual sessions on the Internet. Although the centre does not keep records of users' browsing histories, the manager was of the opinion that the youths "browsed entertainment material," thereby broadening the uses of ICTs among youths.

Also, although the project in Nakaseke is seen to address the question of a rural digital divide, with some positive results, not all the youth in Nakaseke had equal opportunities for ICT use, as contextual issues of poverty (within the household), lack of skills, training, and institutional issues combined to limit their full use of, and participation in, ICT discourses. However, in this era of overwhelming capitalism and materialism, Nakaseke's ability to create communicative spaces of relevance to youths with limited funds is noteworthy.

Conclusions

Based on the fact that youths in Africa constitute a potent force in the continent's development—a fact that legitimates their central role in the processes of democracy and development—the main aim of this chapter has been to investigate the access, participation, and inclusion of youths in rural Uganda in ICT discourses for development.

The findings indicate a bold shift away from active politics to the apolitical engagement necessitated by a dialectical interplay of several factors, including the proliferation of free-market forces that negate conflicting viewpoints, in addition to antecedent contextual influences specific to remote and underdeveloped rural communities such as Nakaseke.

The study also concludes that the agency of ICTs in fostering young people's participation in public discourses through the community media is dictated by broader influences of the structure and operation of the media establishment. The specific linkages between media control, public communication, national electoral politics, and market mechanism are conjectured to weaken political democratic practice at the expense of development.

The argument then is that the divorce of development issues from political debate not only localizes youth discourses but also denies youths their rightful political public sphere, especially because development issues are political and democratic in nature and should be exposed to a market place of ideas.

Notes

1. This chapter is part of Dralega's doctoral thesis entitled "ICT Based Development of Marginal Communities: Participatory Approaches to Communication, Empowerment and Engagement in Rural Uganda," submitted to the University of Oslo in March 2008.
2. Telephone interview with Peter Balaba, the CMC manager, in 2006.
3. For the eight goals, to be achieved by 2015, visit: http://www.undp.org/mdg/basics.shtml. (accessed on September 19, 2007).
4. The Kenyan WSIS Youth Caucus is a consultative forum for young Kenyans and Youth Organizations interested in the participatory development of an Information Society that protects the interests of youths and facilitates sustainable development nationally and globally. They were actively involved in WSIS and ICT Policy process in Kenya (Mihyo and Sesan 2004, 133).
5. Paradigm Initiative Nigeria was an African e-conference held online during November 20–30, 2002, in which ninety-six young Africans from thirteen countries sought to strengthen the role of youths in the WSIS process (Mihyo and Sesan 2004, 133).
6. African Youth Initiative is an African youth-led ICT-for-accelerated-development outfit, representing ten countries, which seeks to bridge the digital divide in Africa and fulfil the desire to use ICT to address the continent's problems while building on the resources of young people (Soltane et al. 2004, 135).
7. The ICT policy provision for license fee exemption was that the radio radius did not exceed 25 km coverage.
8. The CMC manager and some of the interviewees described the violence to the researcher from Nakaseke. It has to be noted that there were similar countrywide acts of violence, some of which was reflected in the national media.
9. Interview with CMC manager in 2006.

References

Acacia. 1998. Acacia (Communities and the Information Society in Africa) Annual Report 1998–1999. IDRC: Ottawa.

Ascroft, J. and Masilela, S. 1994. "Participatory Decision-Making in Third World Development." In *Participatory Communication: Working for Change and Development*, ed. W. K. Shirley and A. Joseph. New Delhi: Sage.

Barbour, S. R. and J. Kitzinger, eds. 2001. *Developing Focus Group Research: Politics, Theory and Practice*. Thousand Oaks: Sage.

Behrouzi, M. 2005. *Democracy as the Political Empowerment of the People: The Betrayal of an Ideal*. Lanham, New York: Lexington Books.

Bell, B. L. 2005. *Children, Youth and Civic (Dis)Engagement: Digital Technology and Citizenship*. CRACIN Working Paper No.5, Concorda University.www.cracin.ca.

Bessette, G. 2004. *Involving the Community: A Guide to Participatory Development Communication*. Ottawa: IDRC.

Brown, B. B., R. Larson, and T. S. Saraswathi, eds. 2002. *The World's Youth: Adolescence in Eight Regions of the World*. New York: Cambridge University Press.

Chibita, M. 2006. Indigenous Language Programming and Citizen Participation in Ugandan Broadcasting: An Exploratory Study. PhD thesis, University of South Africa, Pretoria.

Dervin, B. and R. T. Huesca. 1999. The Participatory Communication for Development Narrative: An Examination of Meta-theoretical Assumptions and their Impacts. In Proceedings, Participation in Communication Systems for Development, ed. J. Diaz-Bordenave, Rio de Janeiro. Unpublished Paper.

ECA. 2006. *Fifth African Development Forum: Youth and Leadership in the 21st Century*. Economic Commission for Africa: Addis Ababa.

Fourie Ina. 2006. *Information Literacy for Healthcare Professionals—Learning from Studies in Human Information Behavior*. Pretoria: University of Pretoria.

Golding, P. and Murdock, G. 1997. *The Political Economy of the Media* (International Library of Studies in Media and Culture, 2). Cheltenham, England: Edward Elgar.

Habermas, J. 1989. *The Structural Transformation of the Public Sphere: An Inquiry into a Category of Bourgeois Society*. Cambridge, MA: MIT Press.

Hagen, I. 1992. *Democratic Communication: Media and Social Participation*. In *Democratic communication in the information Age*, ed. J. Wasco and V. Masco. Toronto: Garamond Press.

ITU. 2007. *Telecommunication/ICT Markets and Trends in Africa* 2007. Geneva: International Telecommunication Union.

Jacobson, T. and S. Kolluri. 1999. "Participatory Communication as Communicative Action." In *Theoretical Approaches to Participatory Communication*, ed. T. Jacobson and J. Servaes. New Jersey: International Association for Media Communication Research (IAMCR).

Jacobson, T. L. and J. Servaes, eds. 1999. *Theoretical Approaches to Participatory Communication*. New Jersey: IAMCR.

Jayaweera, N. 1987. "Rethinking Development Communication: A Holistic View." In *Rethinking development communication*, ed. N. Jayaweera and S. Amunugama. Singapore: Asian Mass Communication Research and Information Centre.

Kinunda, H. 1998. "Kenya: Health Education in Schools—The ECHA Project." In *"Child-to-child" in Africa: Towards an Open Learning*

Strategy, ed. A. K. B. Tay. Paris: UNESCO-UNICEF Cooperative Programme. Digest 29.

MacBride, S. 1980. *Many Voices, One World: International Commission for the Study of Communication Problems.* Paris: UNESCO.

Mann, C. and Stewart, F. 2000. *Internet Communication and Qualitative Research: A Handbook for Research Online.* London: Sage.

Marshall, C. and Rossman, G. B. 1999. *Designing Qualitative Research.* Sage: California.

Mato, D. 1999. "Problem of Social Participation in 'Latin' America in the Age of Globalization: Theoretical and Case-based Consideration for Practitioners and Researchers." In *Theoretical Approaches to Participatory Communication*, ed. T. Jacobson and J. Servaes. New Jersey: IAMCR.

Mbaine, A. 2006. *Media in Situations of Conflict: Roles, Challenges and Responsibility.* Kampala: Fountain Publishers.

Melkote, S. R. and L. H. Steeves. 2003. *Communication for Development in the Third World: Theory and Practice for Empowerment*, 4th edn. London: Sage Publications.

Mihyo, P. and G. Sesan. 2004. "Africa Youth in the Information Society." In *Africa Networking: Development Information ICTs and governance*, ed. B. K. B. B. Soltane, O. N. Fluck, A. Opoku-Mensah, and M. A. M. Salih. Addis Ababa: International Books and ECA.

Mugaju, J. 1999. *Uganda's Age of Reforms: A Critical Overview.* Kampala: Fountain Publishers.

Murdock, G. 2000. "Concentration and Ownership in the Era of Privatization." In

Media Studies: A Reader, 2nd edn, ed. P. Marris and S. Thornhaum. New York: New York University Press.

Mutibwa, P. 1992. *Uganda since Independence: A Story of Unfulfilled Promises.* Kampala: Fountain Publishers.

Mwesige, P. 2004. "Cyber Elites: A Survey of Internet Café Users in Uganda." *Telecommunications: Development in Africa* 21(1): 83–101. Oslo: Norwegian Consortium of Academic Libraries.

Norris, P. 2001. *Digital Divide: Civic Engagement, Information Poverty, and the Internet Worldwide.* Cambridge: Cambridge University Press.

Pateman, C. 1970. *Participation and Democratic Theory.* Cambridge: Cambridge University Press.

Rozumilowicz, B. 2002. *Democratic Change: A Theoretical Perspective.* Routledge Research in Cultural and Media Studies 9: 9–26.

Sarantakos Sotirios. 1998. *Social Research.* London: Macmillan Press Ltd.

Soltane, B. K. B. B., O. N. Fluck, A. Opoku-Mensah, and M. A. M. Salih,

eds. 2004. *Africa Networking: Development Information ICTs and Governance.* Addis Ababa: International Books and ECA.

Tangri, R. 1999. *The Politics of Patronage in Africa: Parastatals, Privatization and Private Enterprise.* Oxford: James Currey.

TI. 2003. *Global Corruption Report: 2003—Regional Reports, East Africa.* Berlin: Transparency International, 237–247.

Uganda Bureau of Statistics. 2007. Entebbe. http://www.ubos.org/

UNDP. 1998. *Human Development Report*. New York: Oxford University Press.

Wanyeki, L. M. 2000. *The Development of Community Media in English Speaking*

West Africa. In *Promoting community media in Africa*, ed. S. T. Kwame Boafo.

Paris: UNESCO.

Yin, R. K. 2003. *Applications of Case Study Research*, 2nd edn. Applied Social Research Methods Series. Vol. 34. New Delhi: Sage.

CHAPTER 9

Transnationalism, the African Diaspora, and the Deterritorialized Politics of the Internet

Wisdom J. Tettey

Introduction

Spatial flows are a primary characteristic of globalization (Castells 1997) and have reshaped not just financial and cultural interactions and transactions, but the development of what Appadurai (1998) calls ethnoscapes, born out of the unprecedented movement of persons across geographically dispersed territories. Intertwined with these mobilities are the possibilities that have opened up for migrants to engage with their places of origin. This involves bringing the imagined communities of "home" to their new locales at the same time as they project themselves onto the realities of the places they have left behind. A corollary to these developments is the phenomenon whereby political, social, and economic actions are not confined to the specific territories of sovereign states, but are breaking down boundaries as transnationals are implicated in the politics and society of their home countries. The emergence of increasingly extensive diaspora communities has, therefore, been attended by strong connections, not dissociation, from their places of origin. These connections have been aided, in no small measure, by information and communication technologies (ICTs). As Tettey (2004, 123) points out, we

> see a new media ecology emerging in tandem with the new physical ecology which is represented by worldwide migrations over international borders, and the formation of diaspora groups.... The convergence of

> these dual processes has created a situation that merits designation as the diaspora of the internet. This refers to the organization of social groups outside their countries of origin as communities of action, not primarily according to the necessities of physical propinquity, but rather by the possibilities presented by the boundlessness of the new technological architecture of the internet.

In the case of the African diaspora, there are several Web sites, Internet TV and radio broadcasts, and other interactive venues, dedicated to providing news and discussions about developments in various African countries. These venues are the main source of information and political engagements for those in the diaspora, helping them to keep abreast with events, issues, and conditions in their home countries at the same time as they provide avenues for civic interaction. Adams and Ghose (2003) define these mediascapes within which transnational conversations take place as a "bridgespace," incorporating not just the Internet but also other electronic media such as music CDs and films. They argue that the concept of the "bridgespace" provides a better analytical frame than "bridgeplace," because the latter connotes a level of coherence that is absent in the media terrain traversed by those who move within transnational circuits. They contend further that the nonphysical geographies of these communicative media are more appropriately captured by the concept of space, which embodies "connections between here and there in both a geographical and cultural sense" (Adams and Ghose 2003, 416).

Drawing on the concept of a bridgespace, this chapter aims at providing critical analyses of the complexities of national imaginings and political engagement that characterize expressions of transnational citizenship among African immigrants in the diaspora who have access to, and participate actively in, the transnational spaces of the Internet. The active participants are also known as core members of the diaspora, and need to be distinguished from passive and silent members (Tettey and Puplampu 2005). The chapter, thus, explores the extent to which Africans in these locales actively utilize their agency, as transnational citizens, within the mediascapes made possible by the Internet, to engage in political discourses about their home countries, pursue long-distance nationalism, and attempt to shape politics and public policy in their countries of origin. The chapter interrogates how the intersecting dynamics of diasporic locations and experiences, and the sociopolitical landscapes of "home" foster expressions of political agency within these online communities and lead to solidarities as well as contestations about various forms of political articulation, mobilization, and participation across the deterritorialized spaces that these Africans occupy.

Transnational Citizenship, Liminality, and Politics in Internet Bridgespaces

ICTs, in general, and cyberspace, in particular, have accentuated the links between home and the diaspora, the local and the global, and have created forums for those connections to be articulated, aggregated, and contested (Androutsopoulos 2007; Atran 2006; Tettey 2004; Tsalikli 2003; Karim 2003; Mitra,1998; Appadurai 1996). Bridgespaces provide opportunities for individuals and groups to exist and extend their networks of support and obligations, linkages that are constitutive of the essence of their being. To underscore the significance of these connections, Adams and Ghose (2003, 41) argue that

> Places are not containers filled with people but are actually "topologies" of relationships between people. The sum of all the one-way and two-way communication links between agents creates place. . . . The topologies in question involve networks of obligation, knowledge sharing and commitment.

The terrains of cyber-interaction are characterized not only by a sense of identification and belonging but also by contestations at different levels of resolution. For people in the diaspora, the contestations are, in some respects, the products of disagreements over issues that have their provenance in the places of origin and have been transposed into the new physical locales as well as the deterritorialized spaces of the Internet. Axel (2004, 45) suggests that diasporas are not just a dispersed group of individuals but an agglomeration of a globally mobile category of identification around which "diverse, local, individual, and present indeterminacies are displaced and qualities of the global, the collective, anteriority, and futurity are invoked and instantiated."

In addition to contestations generated by anteriorities and attendant discursivities, there are others that result from the liminality that characterizes the location of diasporic communities as hybridized individuals and groups. In the "in-between" spaces (Bhabha 1994) occupied by these agents emerges a clash of values, attitudes, perspectives, and allegiances that, at once, might pit them against their host communities and or their places of origin. As Naficy (1993) observes in the context of Iranian immigrants in Los Angeles, migrants engage in a process of exhilic haggling that entails tensions resulting from simultaneous identification with home and an acknowledgment of the difference between home and diasporic locales and experiences. There is, therefore, a need to problematize this liminality in order to appropriately analyze the relationships between home and diaspora, and to adequately understand the nature of politics in the interstitial spaces created by the intersection of home, diaspora, and the Internet.

With respect to transnational politics, participation by diasporic Africans stems from their belief that they are not just unobtrusive observers of, and commentators on, events taking place in a far-off place, but that they are invested in those events and goings on, and are directly or indirectly affected by them. They, therefore, assert a legitimate right and obligation to engage in the politics of home, even if only in the context of Internet sphericules. As Tettey (2004, 129) explains,

> "Place" is at once a physical construct and a mental imaginary. Thus, while people may be separated from the physical construct of "home," as a result of immigration and other forms of geographical mobility, they tend to retain their attachment to that space through mental connections and outward practices that invoke that geographical location.

In her analysis of Internet-enabled political activity by Eritreans in the diaspora, Bernal (2006) argues that this community should be appropriately conceptualized as an "offshore citizenry." The community, which is characterized by what she calls "emotional citizenship," is able to take advantage of its location to facilitate certain purposes that affect the imagined "home," which, though symbolized by the physical territory of the state, is not limited by it. Thus, "what might have once been outside the margins (of the nation) is now effectively included within a larger framework of imagined community" (Bernal 2006, 163). The extent to which Africans in the diaspora are emotionally invested in home politics is illustrated by the extensive postings in chat-rooms among South Africans during, and following, the race for President of the African National Congress. One contributor contends that

> With the credentials that zuma has. [*sic*] It is nothing more than a very poor joke that he will lead this country. He has the ability to turn this country into a second zimbabwe. He will undo all the good work done by Nelson Mandela and the rest of the honest anc. members. pity but I think rather than see this beautiful country go the way of the rest of africa it may be time to consider immigrating (Furrows 2007).

The sentiments expressed above reflect the disdain among many South Africans in the diaspora, and indeed the view of many of their compatriots at home, that Zuma's image has suffered as a result of his rape trial; statements that he made during that trial about Zulu male sexuality and HIV/AIDS; as well as charges of complicity in an arms scandal for which he was being prosecuted. They are of the view that that image diminishes Zuma's stature on the global scene and that, by extension, allowing him to represent the country as president will be a blemish on their homeland's standing in the world—a standing that was very positively cultivated by Mandela following apartheid. Some of the negative views about Zuma flow from discomfort with his more populist and leftist orientation to politics. This orientation breeds anxieties, among his ideological rivals, about state takeover of the economy and a "revolution from below" that could parallel what has happened in Zimbabwe, with its devastating consequences.

The quote below, generated in the context of charges brought against the speaker of the Nigerian House of Representatives also captures the extent to which people in the diaspora feel intricately connected to developments at home.

> When it comes to Madam Etteh herself, even if we give her the benefit of the doubt and buy her story that she was not part of the process of deciding what needed to be done nor the contractor to do the job, she is at the very least guilty of immorrality and a lack of compassion. For how else could she approve 5 million dollars for the renovation of two homes which by the way I have seen and will at the most cost 25 million naira to build from scratch. If she cannot see why it is completly immoral and unacceptable for her to spend that sort of money on herself while those she is suppossed to serve wallow in poverty, then she does not deserve to be speaker (Shatu, UK, Posted: Wed Sep 26, 2007 2:05 pm, http://www.nigeriaworld.com/board/viewtopic.php?t=3823)

The significance of the foregoing discussion for anthropological analyses of the local is significant. We need to acknowledge the extent to which the local is inflected in processes of globalization and vice-versa, thereby requiring a radical rethinking of the privileged position that the local has assumed in anthropology. This involves focusing on

> the historically specific ways that localities tend to indicate a world beyond themselves, an indication around which, indeed, new sentiments of longing and belonging may be generated. Ethnography may begin to understand the local dialectically as translocality. By this, I mean not merely that we may study how empirical localities are connected to distant places through various communications technology but that we may develop ways to study how emergent processes of globalization are productive of locality (Axel 2004, 47).

A very important consequence of diasporic location and the opportunities provided by the Internet is the extent to which the state's hold on its translocated citizens has been ruptured, if not completely eliminated. Under the media configurations of the past, characterized by centralized control of mass-mediated information production and dissemination in African countries, the ability of citizens to produce content and share their views was constrained by the gatekeeping role of state agents. The transformations in media ownership, control, and information dissemination that has followed democratization in many parts of the continent over the last several years, resulting in the expansion of private media outlets, combined with the ability of these media outlets, as well as individual citizens, to bypass state control, implies that counter-discourses that challenge the hegemonic viewpoint of the state are possible and thriving in the transnational bridgespace provided by the Internet. Chan (2005) notes that in the past resistance from the audience was confined to the private realm, "because traditional mass media operate by the logic of 'few producers, many receivers,' few members of audience have creative access to the media to produce their own counter-discourses.... Internet users, on the other hand, can produce and disseminate their own content and views" (see also Mitra 2001). Cyberspace can, thus, become a discursive space for multivocality.

These opportunities are particularly available to Africans located outside the territorial boundaries of their states of origin. The location of Africans in the diaspora affords them the opportunity to access and assess information in ways that most of their compatriots may not be able to. They are, consequently, better able to authenticate or debunk various narratives that may be put out by state actors in their countries of origin. These are then posted to support or challenge those dominant discourses. Discussion forums and chat-rooms, thus, provide a plethora of deliberative politics, as burning issues are brought up, opinions expressed, and analyses provided. This is illustrated by Bernal's (2006) study of Eritrean Internet networks that provide a veritable public sphere in which ideas, which cannot be expressed in the context of the oppressive controls of the state, are articulated, disseminated, and debated among interactants who may not be familiar with each other in physical space but engage one another in the ether of cyberspace. It is obvious that some posters (such as Nigeria's Oladipo in Tettey 2008, 10) would not have the temerity to say or write the same words in the real world of African politics without fear of physically harmful retribution.

One of the visible outcomes of the Internet public sphere is the opportunity it has created for public intellectuals to emerge and share their thoughts on burning political issues. Some of these individuals have provided professional perspectives that may not elicit agreement from all interactants, but are nevertheless respected. However, the potential of the Internet to provide a space for genuine deliberative politics referred to above is constrained by the fact that discussions are still characterized by some of the unsavory interactions that prompted Tettey (2001, 146) to conclude that

> The discussions... essentially do not contribute much to advancing a discourse that promotes democratic ideals.... By and large, discussions

> degenerate to name calling and vituperative partisanship that appeals to ethnicity in most cases…the forums are becoming spaces where demagoguery stifles the pursuit of democratic ideals.…Instead of offering a space for rational political discourse, the forums seem to spawn irrational emotion.

It is also critical, as we analyze Internet-enabled politics within the African diaspora, that we do not infuse that space with the logic of unbridled "democratic equivalence" (Cunningham 2001, 132) for all participants. Rather,

> the very nature of the discussions eliminates certain categories of citizens from participating with the same level of authority and equality that is accorded others who are considered "qualified" to voice their opinions on matters under discussion…(Tettey 2001, 147).

The liminality of translocational positionality (Anthias 2001) and the relative communicative autonomy provided by Internet bridgespaces produces a dialectical tension between subject and citizen, which is significant for negotiating interactions between Africans in the diaspora and their states of origin. Glazer (2002) acknowledges the perennial conflicts that result from dual citizenship, for example, as migrants negotiate their loyalty to host and home countries. "Increasingly, states can no longer assume exclusive loyalty, and divided or competing loyalties derive from both migrants and states alike" (Brainard and Brinkerhoff 2006). A corollary to the subject-citizen dialectic is, therefore, the delimitation of state sovereignty vis-à-vis the diaspora, thereby contradicting the assertion of state domination in recent literature on the state (Mbembe 2003; Hansen and Stepputat 2005). The activities of diaspora Africans in cyberspace and the influence these have on home state politics shows that citizens are not simply subject to state power but can be both outside of it, during particular performative moments, and constitutive of it—shaping its actions and discourses.

While the exact impact of politics within the diaspora of the Internet cannot be definitely ascertained, and a causal linkage cannot always be established between specific deliberations and postings, on one hand, and political actions at home, on the other, they do influence domestic politics. There are several instances where issues that have been raised within chat-rooms or on Web sites, principally by self-identified members of the diaspora, have found their way into mainstream national politics—influencing policy, eliciting a response from state officials, or providing fodder for domestic political actors of one kind or another. In fact, state officials monitor discussions and postings on the Internet in order to apprise themselves of issues that have implications for domestic politics, ascertain responses to government policies, and so on. There are even speculations that state officials participate in Internet forums and chat-rooms in order to have a handle on issues of interest to the diaspora community or influence discussions. Bernal (2006,173) suggests, in the context of an Eritrean Internet site, that "since readers cannot always tell if a poster is using their [*sic*] real name or a pseudonym, some readers suspect that members of government pose as ordinary posters to push online debates in pro-government directions." Previous research indicates that this is not mere speculation, and that at times state officials do respond in their official capacities—sometimes without anonymity (Tettey, 2001). In fact,

> since 2006, the Ethiopian government has blocked access to websites and blogs that it considers critical of the regime, with the Ethiopian Telecommunications Agency instructing cybercafés to record names and addresses of customers so that they could be punished for any untoward behavior, including imprisonment. Similar requirements obtain in Tunisia. Zimbabwe has secured Chinese technology to enable it implement the provisions of the Interception of Communication Act, which authorizes state agencies to monitor telephone, fax, and e-mail messages. The Egyptian government, on its part, has increased online censorship, coming down very hard on bloggers who express critical views.... in the summer of 2006, the government passed a law on internet regulation that facilitates such clamp-downs under the pretext of protecting national security (Tettey 2008, 14).

As a result of such monitoring, a U.S.-based Gambian journalist, Fatou Jaw Manneh, was arrested, in March 2007, when she arrived at Banjul airport, because of her postings on the Internet.

> [She] was prosecuted for an article, in October 2005, in which she accused President Yahya Jammeh of "tearing our beloved country to shreds" and describing the head of state as a "bundle of terror." She was charged with "intention to commit sedition," "publication of seditious words" and "publication of false news intended to create public fear and alarm" and faces three years in prison (Reporters Without Borders 2008, 17).

From the above, it is obvious that cyberspace has encouraged the development of a civic culture and an active citizenry that integrates the local and the global, and compels the state of origin to be responsive to a public sphere that is external to it and over which it does not exercise complete sovereign control. This is not to suggest that domestic politics is driven by virtual politics, but just that it is inflected by it in some instances. As pointed out by members of the Internet site Dehai, the forum has "influenced political policies in Eritrea...Although the government does not officially respond to statements on Dehai, sometimes after a criticism is posted, the policy is changed" (Bernal 2006, 173).

Post-National Constellations and the Politics of Difference in Internet Sphericules

Fraser (2005, 1) critiques the national/Westphalian orientation of public sphere theory, which she thinks is threatened by depoliticization, and calls for a reformulation of the theory in a way that "can illuminate the emancipatory possibilities of the present postnational constellation" associated with globalization, postcoloniality, and multiculturalism. In the new post-national constellation, the actors, and the communicative targets of those who operate within the public sphere, contra the Habermasian model of the bourgeois Westphalian state and public, are not just the bourgeoisie and the territorial sovereign. At work are multiple actors and a desegregated sovereign, characterized by multiple and distinct institutional levels—some contained within, and others extending beyond, the state. As Africans engage in communicative action on the Internet, they manifest these permutations and characteristics of the public sphere. Therefore, the actors in the virtual public sphere and the institutional and other receptacles, to which they are directing attention, reflect these multiple foci and loci of sovereignty and action, depending on the issues at hand. It is worth noting that they are not functioning in a conceptually unique African public sphere. Rather, what their activities tell us is the need to recognize an understanding of the public that is not constrained by the Habermasian model and incorporates the sophisticated landscape of political engagement that we see in Africa and elsewhere.

Tettey and Puplampu (2005, 164) point out that "transnationalization of national politics and the diasporization of civil society provide a means whereby the African diaspora participates in, and influences, the domestic politics of their countries of origin." Such engagement, in relation to political parties and political propaganda, has been significantly transformed by the use of the Internet. The technology has not only allowed opponents and supporters of political parties, in the diaspora, to widely circulate propaganda and mobilize resources but has also made it possible for politicians and political parties at home to appeal directly to members of the diaspora. The U.S. Branch of Ghana's New Patriotic Party, for example, has a Web site (http://nppusa.org/) devoted to disseminating the party's accomplishments, showcasing its candidates, and seeking donations from members of the diaspora to enhance the party's fortunes in elections. In the same vein, some individual candidates for the country's various political parties' presidential slot in the 2008 elections set up Web sites to reach out to voters. Some of these Web sites provided not only textual but audiovisual content as well (for example, www.attamills2008.com; www.nduom.com; www.osafo-maafo.com). A significant constituency that is the target of these electronic activities is the Ghanaian diaspora, which comprises potentially major contributors to the finances of these candidates. The extent to which politicians travel abroad to court the support of their compatriots testifies to the value of this constituency for domestic politics, particularly when it comes to the mobilization of funds (see Tettey and Puplampu, 2005). Furthermore, the fact that several options for financial donations are not easily available to people in Africa (for instance, use of credit cards and *Paypal*) proves that the diaspora is the primary target of such solicitations. Overall, the level, extent, and tone of political discussions, in African forums, increases during election seasons in various countries, with no shortage of partisan opinion on politicians, political groups, and outcomes of the polls.

While much of the politicized spaces created by the diaspora of the Internet have to do with domestic politics in the countries of origin, we should not lose sight of the extent to which they also provide a means for engaging with entities outside of the territorialized construct of home states, but whose actions or views impact the deterritorialized imaginary, of which the diaspora is a constitutive element. Thus, participants in these spaces engage in counter-hegemonic discourses vis-à-vis what they perceive to be ideological or value impositions by global forces that are inimical to the interest of their societies. In response to criticisms of the Mugabe regime in Zimbabwe, for example, a contributor asserts that

> Mugabe has stopped European terrorists from operating in Africa, Mugabe has stopped Europeans from controlling majority of Zimbabwean land. He is the only African leader to have addressed properly, the land reform issues. This is why European criminals and their medias continue to try to bring Mugabe and Zimbabwe down through many means especially economic terrorism as well as trying other forms of sabotage.... (Sniper, Ireland, http://www.topix.com/forum/world/zimbabwe/T9IIEGU8L1M07CVMK)

The notion of counter-hegemony, particularly as it pertains to political issues, is not without contestation either, as other voices articulate positions that suggest that at a different level of resolution, these supposedly counter-hegemonic viewpoints are themselves reflective of an attempt to impose a singular and dominant frame of values within Internet spaces as well as in state policy. The dominant discourses are therefore relative, and the challenges expressed against each version emphasizes the fact that there is no single unifying meta-narrative in the bridgespaces constituted by the diaspora of the Internet. This is evidenced by the following retort to the quote above in support of Mugabe:

> Mugabe's regime is nothing more than a collection of criminals and misfits since Idi Amin and Pol Pot, why you clowns are defending him is beyond me. The US and Britain should have invaded the place in 2003 instead of Iraq (Timemachinesalesman, Clarkson, Canada).

One of the most passionate forms of political discourse and mobilization via the Internet has to do with what Anderson (1998) refers to as long-distance nationalism. The technology has been employed by various nationalism projects to champion, and mobilize, support for their cause (see Bakker 2001; Tekwani 2003; Kluver and Qiu 2003). Citing an example from the Chinese diaspora, Chan (2005, 342) notes that

> Contrary to Turkle's (1995) thesis that identity on the internet is more fluid and fragmented, Sun (2002) found that the transnational imaginaries of Chinese migrants from the PRC were still rooted in the discourse of a historicized nation, and anchored to the bounded territoriality of a sovereign state...

Analyses of diasporic African spaces on the Internet echo similar findings. Bernal (2004; 2006) provides copious information to show that members of the Eritrean diaspora have used the technology in significant ways to articulate their sense of belonging to their homeland, aggregate support for its nationalist struggles vis-à-vis Ethiopia, organize demonstrations in respect of that cause, and strongly advocate its point of view to the international community in order to court sympathy for that position. She notes, further, that beyond advocacy and political mobilization, this particular group also used the Internet to raise funds in support of the Eritrean government in the 1998–2000 border war against Ethiopia (Bernal 2004; 2006).

The vituperative nature of nationalist discourse among members of the diaspora is substantiated by the following exchange between an Ethiopian and Eritrean:

> hahhaaahahahahahaaaahahahahaha hahahaaaahahaha...look who is talking, eritrea is the largest food aid and donor recipent country in the whold world and 2/3 of it's people (which accounts for 66.6%) are in dire need of food aid and you turn around talk crap? eritrean man (mega stupid) in case you failed to attend school to understand the current situation; you can always ask people to explain to you...by the way what did stop you from taking BADME? IF YOU ARE A MAN

> LIKE YOU BLAHBER ON THE WEB SITES WHY DON'T YOU GO TAKE BADME?…OH, SORRY I FORGOT YOU ARE NOT MAN ENOUGH. BADME IS UNDER ETHIOPIAN CONTROL SO FAR. WHAT YOU GONNA DO ABOUT IT? ERITREANS (Debol, Portland, ME, http://www.topix.com/forum/world/somalia/T0CRSCMC8JUM3VTU5/p75)

Such an incendiary character to some of the nationalist exchanges among members of the diaspora has compelled some of their compatriots, who have to live with the consequences of such rhetoric *in situ*, to castigate those whom they consider to be stoking the fire of nationalist conflict. In a posting, corroborating Anderson's charge that diaspora nationalists are unaccountable and irresponsible (cited in Lyons 2003, 5), an Ethiopian resident laments thus:

> Most of you packed your bag went to the western world to enjoy the freedom they preserved for such a long time.…But what are you doing to the people of the horn, litting fire, sprinkle benzene so that we, who are in the region, burn and dimnished to ash (Elilta, Saturday Dec 15 http://www.topix.com/forum/world/somalia/T0CRSCMC8JUM3VTU5/p77).

The defense of national identity and espousal of national pride is not only manifested in situations of armed conflict or interstate tensions. Discussions in Internet chat-rooms and forums suggest that members of the diaspora community, while critical of their governments and the actions of their compatriots, are inclined to bind together to defend their common identity, as members of a particular country, against perceived negative perceptions by others, even in seemingly mundane situations. In this relational context, they tend to draw the distinction between the actions of a few, which might elicit negative comments, and what they present as a generally good collective of people who share that identity. Nigerians on the *Nigeriaworld* message board were, for example, incensed by postings elsewhere that castigated their collective identity and portrayed them in unflattering terms.

> Most Africans accuse us of aggression when we are merely being assertive. They claim we are materialistic whereas the truth is that we demand the best always.…On balance Nigerians do far more good than harm but in a world where a man cannot even recognise his own inferiority complex it is easier for other Africans to villify Nigerians when they truly envy and covet their abilities, charisma and Natural flair (Oriade, http://www.nigeriaworld.com/board/viewtopic.php?p=25496&sid=48d04070ec0dc87a35ddd431ef13c8a5#25496).

In spite of the general disdain for negative portrayals of themselves and their compatriots, some members of the forum were more sophisticated in their assessment of the reasons their collective identity is assailed and tarnished. This is exemplified by the following post:

> Gentlemen, it is true that Nigerians are being victimised unjustly all over... Having said all these, it is also true that much of the stigma from Nigerians' criminal ways of life partly justify these global blacklisting of Nigerian citizens.... As we condemn worldwide discrimination against Nigerians, we must also examine our behaviors too (http://www.nigeriaworld.com/board/viewtopic.php?p=25502&sid=48d04070ec0dc87a35ddd431ef13c8a5#25502).

What can be surmised from this response to the first posting is the fact that nationalist solidarity among diaspora Africans is not based on an ossified and uncritical commitment but rather subject to introspection and context. As Chan (2005, 75) observes, although the Internet may be characterized by centripetal forces that draw migrants together, "there is, on the other hand, a powerful segmenting or centrifugal force (see also Dahan and Sheffer, 2001)... produced by the multiple and varied discourses on the image of the nation, which reflect the inner differences and contradictions within the virtual community." The following exchanges among Nigerians, on the occasion of the country's forty-seventh anniversary of independence, are instructive. One version of the Nigerian nation proudly states:

> Long live the greatest country on earth. May we be free from the latent stranglehold that the colonialist continue to apply on our people, economy, mineral and human resources. Lets not forget that we are the brightest of the bright, the envy of peoples black and white. May we remain the happiest of all people in the world. Long live ONE Nigeria! (Musa, Posted: Sat, Sep 29, 2007 10:29 PM)

This image is strongly challenged by another contributor who admonishes that

> Nigerians should be wiling [*sic*]. to be honest at least for once, and we human beings often find it impossible to separate fact from fiction. To reiterate that all the statements made above is [*sic*] complete fallacy is to emphasize the sparkling obvious (Mike 57 Posted: Sun, Sep 30, 2007 3:27 AM).

What the above exchange shows is that whereas some in the diaspora feel a sense of pride toward their countries of origin, there is, at best, cynicism and ambivalence among some Africans in the diaspora about whether they can be justifiably proud of their home countries. The cynicism and ambivalence is reflective of their disappointment in the countries' achievements and, even more especially, the political leaders.

In addition to the aggregation of subjectivities that focus on building solidarity for a state in the Westphalian sense, Internet-facilitated long-distance nationalism is also manifested in relation to subnational entities such as regional, racial, linguistic, or ethnic groupings. Indeed, a significant part of the political discourses that engulf cyberspace pertain to ethnic-nationalism, which, by extension, elicits interethnic strife within diasporic communities. For a lot of these interactants, the Internet not only provides an avenue to contest others' rendition of national and group histories (Ho, Babeer, and Khondker 2002); it also makes possible an avenue to explicate their own versions, foster a community of affinity behind their group causes, and mobilize for political action or deliberation. These uses of Internet chat-rooms, Web sites, and listservs are reflected in activities on various African sites. These activities resonate a view expressed by Bhabha (1994, 145–148) to the effect that cyberspace has become "a liminal signifying space that is internally marked by the discourses of minorities, the heterogeneous histories of contending peoples, antagonistic authorities and tense location of cultural difference." The veracity of this observation is borne out not only by contestations of national narratives but also by subnational narratives that embody affinities with, and appeals to, ethnicity or other forms of partisan politics.

A significant dimension of the post-national constellation is epitomized by the contestations regarding the national interest and the history of the nation, concepts that were homogenized under the Westphalian configuration and around which a definitively identifiable and united community of solidarity could, presumably, be mobilized. It is in this respect that Fraser's (1992) focus on "subaltern counterpublics" is significant. These groups, which operate on the margins of the dominant public sphere are repositories of resistance discourse in the spaces provided by the Internet as they seek to delegitimize hegemonic positions on a variety of issues. Cunningham (2001, 134) identifies several such sphericules, which "might stand as a model based on culture, identity and voluntary belonging rather than based on rights derived from, and obligations to, a state." In an argument about religion, ethnic politics, and national unity on a Nigerian forum, one contributor illustrated these issues by suggesting that

> Nepotism, tribalism and mutual tribal mistrust continues to work to the detriment of this failed contraption we call Nigeria everyday !!!! . . . as president of Nigeria his duty is to seek the development of the whole of Nigeria and not the North alone. . . . Yar Adua's [*sic*] appointments so far

> have had an obvious Northern skew to it when better qualified Southerners exist. I am sure that the Yoruba and hausa elite have continously schemed the Igbos out of the highest office in the land because of their fear that an Igbo president will begin the process for the legal seperation of Nigeria. Since Biafra, the North and reprehensible, treacherous Yoruba leaders like OBJ have decided to freeze out the Igbo man (Oriade, Posted: Wed, Sep 26, 2007 7:47 AM).

The racialization of diaspora South African discourse in cyberspace is also revealing of the different definitions of the nation and the tensions that still beleaguer relations not only between whites and blacks in the postapartheid period but among different groups of whites as well. A posting by a self-described white person, suggesting that black South Africans had legitimate claim to the country's lands, elicited strident rebuff from other whites, which, in turn, evoked counter-narratives about who can lay legitimate claim to the country and its accomplishments. The tenor of the discourses is represented by the following examples:

1. This thread reeks of stupidity. South Africa belonged to the Bushmen, who no longer exist. The current blacks are from the north. South Africa does not belong to them. To the thread starter: I seriously have never been so angry behind this computer.... it was whites who have made such a huge contribution to South Africa by building Cape Town and other wonderful cities (Bowiefan 871 http://www.southafrica.com/forums/south-african-politics/5824-africa-belongs-black-people-4.html).
2. Back the truck up.... Now... oh boy. "Bushmen" is a derogatory term coined by settlers; the San and Khoi Khoi people very definitely do exist even today,... Finally, everybody built Cape Town. Proportionally very few white hands toiled brick-for-brick putting the cities of South Africa together. You will never win a moral argument while saying that good things came out of apartheid and the segregation... (Ches, http://www.southafrica.com/forums/south-african-politics/5824-africa-belongs-black-people-4.html).

Memory, remembrance, and memorialization, and their subjective renditions are a significant part of the politics of the Internet. These are evoked not only through textual representations and construction of national and subnational histories in Internet spaces, as represented by the South African case above, but also through the evocative, and sometimes provocative, images that are posted and disseminated. In a discussion of the intersections among the Internet, state-sponsored terror, and the subjectificaton of subjugation and martyrdom through imagery, Axel (2004, 35) notes that

> Khalistani activists began procuring photo images of tortured bodies of Sikh men (valorized as martyrs) and circulating these on the Internet along with survivor testimonies and historical narratives of the Sikh nation (qaum).... By generating an archive-indeed, a cyber-archive-of Sikh struggles, Khalistanis have established an extremely effective set of practices that remake the putatively private scene of national torture into a transnational spectacle of subjectification. Concurrently, the iconic image of the pure and unscathed amritdhari Sikh has been coupled with another image, that of the tortured body.

The ability of the Internet to allow for subjective narratives of history is significant enough; but its capacity to bring alive images that are ubiquitous, accessible to a myriad of people, (re)generate passions, and contest particular versions of history, makes it a very powerful tool for identity politics of various sorts. In the African context, such imagery is exemplified by a Web site that was set up by some citizens of Anlo, in the Volta region of Ghana, to publish images of victims of what they considered to be government-supported police brutality during a confrontation among factions in a chieftaincy dispute. The caption accompanying the images, states:

> This Website (prototype) is being created to Honor of [*sic*] the Heroes and those who Paid the Ultimate Sacrifice to protect The Heritage and Culture of the Anlo's [*sic*] on November 1, 2007 in the Volta Region of Ghana. May Their Souls Rest in Peace! (http://anlostate.org/)

Another example is captured by comments in reaction to the posting of a video clip, on the Biafra war, on a Nigerian discussion forum (http://youtube.com/results?search_query=biafran+war&search=Search). The reactions encapsulate key issues discussed in the preceding paragraphs and are worth quoting *in extensio*:

> Thank you Afroman007 for posting this. Some of us who were born after the war are being fed with false information...(appollopd).
>
> I have been watching this video...and have also been crying....I was about 15 years old then...I am heart broken (18battalion).
>
> 18battalion,...I am more than heart-broken...I have crying too....our fathers fought to save us....I am proud of my fathers (appollopd).
>
> All Igbo must stand up together and Support the non-violent self-determination...(Okoroman).
>
> Dear Okoroman. Am against non-violence...I believe in violence...(myu2404).

The insertion of the global into local politics and the diasporization of domestic politics have been given a strong fillip by recent developments in international politics, which place limits on previous notions of state sovereignty. Concomitant with these limits is the expansion of the responsibilities of the international community to protect vulnerable populations if there is a "responsibility deficit—if the state proves unable or unwilling to protect citizens, or itself becomes the perpetrator of violence against its own citizens" (International Commission on Intervention and State Sovereignty 2001, 41; See also Brainard and Brinkerhoff 2006). Therefore, as we interrogate the power of the visual in articulations of subalternity and oppression, we need to keep in mind the point made earlier about the disaggregated sovereign and the extent to which its various permutations, some extraterritorial in location, may be the target of appeals for solidarity and support in intrastate conflicts.

While the textual and visual grammar of positionality by various actors is a powerful tool in the discourse of representation and resistance, and the search for solidarity, it is important that we attend to them with caution. This is because of the tendency for the passion of nationalism to take precedence over "truth" in some instances, and the multiple "truths" that are propagated in relation to this phenomenon. It is, therefore, important to bear in mind the following admonishment by Machiavelli (1950), particularly when we analyze exhilic discourses:

> How dangerous it is to trust the representatives of exiles...such is their extreme desire to return to their homes that they naturally believe many things that are not true, and add many others on purpose....A prince therefore should be slow in undertaking any enterprise upon the representations of exiles, for he will generally gain nothing by it but shame and serious injury.

This is not to suggest that the nationalist causes represented by exiles are without merit or that we should be cynical about their claims and assertions. What is being argued is that the Internet provides an avenue for articulation of different truths, which, while controvertible, may not always lend themselves to easy verification. This is because they are not always juxtaposed spatiotemporally with counter-narratives and, even if discredited, are still likely to be accessible in the public realm of the Internet. Consequently, they maintain the potential to be seen as reflecting reality, at least by some, and hence fester divisions on the basis of information that is not always, or necessarily, reflective of reality. Androutsopoulos' (2007, 521) observations are apposite when he says "the views expressed are [not] always unanimous or unprejudiced."

Ethnic politics within the diaspora of the Internet is very closely related to the politics of language, as evidenced by analyses of the sociolinguistic ecology of interactional spaces of the technology. Code-switching, for example, is not only employed to speak to those with whom one shares a language but is also emblematic of sociolinguistic assertion, representation, resistance, and or hierarchization, which are manifested in discursive constructions vis-à-vis others, thereby indexing boundaries among interactants. In the words of Auer (2005, 409), "language alternation can be void of identity-relevant meaning in some contexts, yet in others extremely rich in the identity work it accomplishes." Thus, while the principal language of communication in the African forums and Web sites are the official languages of the countries concerned, participants tend to insert their own vernaculars into conversations in order to assert the authenticity of their attachments, evoke particular sentiments, and situate themselves vis-à-vis others.

Conclusion

This chapter shows that many Africans in the diaspora exhibit very strong emotional citizenship toward their countries of origin. Admittedly, such attachments precede the advent of the Internet. However, the technology has enhanced such connections and provided a bridgespace for articulating a variety of positions toward the imaginary construct they call home. As Bernal (2006, 176) points out, cyberspace reflects

> the struggles of ordinary people to participate in national debates, narrate history, define legitimacy and articulate a moral order. Eritrean websites have fostered the emergence of counter-publics and spaces of dissent where unofficial views are voiced and alternative knowledges are produced. These spaces of creativity at the margins are perhaps all the more important given the pervasive reach of global capital, media conglomerates and regulatory authorities of various kinds.

It is clear from the analyses that while the Internet can facilitate group aggregation, it can also provide the incubus for fragmentation along various markers of difference. The case of the diaspora of the Internet corroborates findings by Adams and Ghose (2003, 415) who aver that

> ironically, when "place-transcending" technologies facilitate the creation of ties through space and reduce the separation between here and there, negating place, this can strengthen a sense of ethnic identity, which implies a tie between self and place. A further irony is that a new technology can be used to cognitively connect with what is a symbol of primordial essence.

Thus, contrary to the tendency to equate technological innovations to progress, the evidence adduced in this chapter indicates that these technologies could be appropriated for purposes that can arguably be considered conservative, at best, and retrogressive, in some instances.

The findings also suggest that the liminal space occupied by the African diaspora shapes their subjective positions and the phenomenological and hermeneutic lenses from which some of the discourses analyzed above are performed. For some people, and in some instances, these spaces have provided a context that has enabled them to contest, for example, received notions around national politics, while, for others, they have enabled appeals to solidarity and reinforcement of commitments to home that come with the vicissitudes of life abroad and determination to improve conditions at home. Based on the preceding discussion about the dialectics between state and substate levels of affiliation/alienation and the solidarities/fissures that they spawn, it can also be concluded that Africans in the diaspora tend to oscillate between essentialized, homogenized notions of belonging and attachment to both a mental and physical construct of home, on one hand, and diverse, multiple, sometimes contradictory and sometimes divisive, positionalities, on the other. Irrespective of where they stand at particular performative moments, however, one thing is clear. They are taking advantage of the opportunities provided by Internet technology to engage unabatedly with the politics of places they call home. In a nutshell, "cyberspace may have annihilated distance but not place" (Walmsley 2000, 5).

References

Adams, P. C. and R. Ghose. 2003. "India.com: The Construction of a Space Between," *Progress in Human Geography* 27(4): 414–437.

Anderson, B. 1998. *The Spectre of Comparisons: Nationalism, Southeast Asia and the World*. London: Verso.

Androutsopoulos, J. 2006. "Multilingualism, Diaspora, and the Internet: Codes and Identities on German-based Diaspora Websites." *Journal of Sociolinguistics* 10(4): 520–547.

Anthias, F. 2001. "New Hybridities, Old Concepts: The Limits of 'Culture.' " *Ethnic and Racial Studies* 24(4): 619–641.

Appadurai, A. 1996. *Modernity at Large. Cultural Dimensions of Globalization*. Minneapolis, Minnesota: University of Minnesota Press.

Atran, S. 2006. "A Failure of Imagination (Intelligence, WMDs, and 'Virtual Jihad')." *Studies in Conflict & Terrorism* 29: 285–300.

Auer, P. 2005. "A Postscript: Code-Switching and Social Identity." *Journal of Pragmatics* 37: 403–410.

Axel, B. S. 2004. "The Context of Diaspora." *Cultural Anthropology* 19(1): 26–60.

Bernal, V. 2004. "Eritrea Goes Global: Reflections on Nationalism in a Transnational Era." *Cultural Anthropology* 19(1): 3–25.

Bernal, V. 2006. "Diaspora, Cyberspace and Political Imagination: The Eritrean Diaspora Online." *Global Networks* 6(2): 161–179.

Bhabha, H. K. 1994. *The Location of Culture.* London: Routledge.

Brainard, L. A. and Jennifer M. Brinkerhoff. 2006. "Sovereignty Under Siege, or a Circuitous Path for Strengthening the State? Digital Diasporas and Human Rights." *International Journal of Public Administration* 29: 595–618.

Castells, M. 1997. *The Power of Identity.* Oxford, UK: Blackwell.

Chan, B. 2005. "Imagining the Homeland: The Internet and Diasporic Discourse of Nationalism." *Journal of Communication Inquiry* 29(4): 336–368.

Cunningham, S. 2001. "Popular Media as Public 'Sphericules' for Diasporic Communities." *International Journal of Cultural Studies* 4(2): 131–147.

Dahan, M. and G. Scheffer. 2001. "Ethnic Groups and Distance-Shrinking Technologies." *Nationalism and Ethnic Politics* 7(1): 85–107.

Fraser, N. 1992. "Rethinking the Public Sphere." In *Habermas and the Public Sphere*, ed. C. Calhoun. Cambridge, MA: MIT Press.

———. 2005. "Transnationalizing the Public Sphere." http://www.republicart.net/disc/publicum/fraser01_en.pdf.

Furrows, N. 2007. *Re: Zuma? God Help Us!* ("Reply #1 on: November 29, 2007, 08:15:03 AM" http://www.iafrica.com/php-bin/forums/NEWS/index.php?topic=919.0).

Glazer, N. 2002. "Dual Citizenship as a Challenge to Sovereignty." In *How Governments Respond: Sovereignty Under Challenge*, ed. J. D. Montgomery and N. Glazer, 33–54. New Brunswick and London: Transaction Publishers.

Hansen, T. B. and F. Stepputat, eds. 2005. *Sovereign Bodies: Citizens, Migrants and States in the Postcolonial World.* Princeton: Princeton University Press.

Ho, K. C., Z. Babeer, and H. Khondker. 2002. "'Sites' of Resistance: Alternative Websites and State-Society Relations." *British Journal of Sociology* 53(1): 127–148.

International Commission on Intervention and State Sovereignty. 2001. *Report of the International Commission on Intervention and State Sovereignty: The Responsibility to Protect.* Ottawa: IDRC.

Karim, K. H. 2003. "Mapping Diasporic Mediascapes." In *The Media of Diaspora*, ed. K. H. Karim, 1–17. Oxford, UK: Routledge.

Kluver, R. and J. L. Qiu. 2003. "China, the Internet, and Democracy." In *Internet and Democracy in Asia: Rhetoric and Reality*, ed. I. Banerjee, 26–60. Singapore: Times Academic Press and the Asian Media and Information Centre.

Lyons, T. 2003. Note on Globalization, Diasporas, and Conflict. http://intlstudies.ucsd.edu/iicas/pdf/lyons.pdf (accessed March 17, 2003).

Machiavelli, N. 1950. *The Prince and the Discourses* (Chapter 31). New York: Random House.

Mbembe, A. 2003. “Necropolitics.” *Public Culture* 15: 11–40. <“Mbembe” is correct>.

Mitra, A. 1998. “Virtual Commonality: Looking for India on the Internet.” In *Virtual Culture*, ed. S. G. Jones, 55–79. London: Sage.

Mitra, A. 2001. “Diasporic Voices in Cyberspace.” *New Media and Society* 3(1): 29–48.

Naficy, H. 1993. *The Making of Exile Cultures: Iranian Television in Los Angeles*. Minneapolis: University of Minnesota Press.

Reporters without Borders. 2008. *2008 Annual Report—Africa*. Paris: Reporters without Borders.

Tekwani, S. 2003. “The Tamil Diaspora, Tamil Militancy and the Internet.” In *Asia.com: Asia encounters the* Internet, ed. K. C. Ho, R. Kluver, and C. C. Yang, 175–192. London: Routledge Curzon.

Tettey, W. J. 2001. “Information Technology and Democratic Participation in Africa.” *Journal of Asian and African Studies* 36(1): 133–153.

———. 2004. “Globalization, Diasporization and Cyber-Communities: Exploring African Trans-Nationalisms.” In *Globalization and the Human Factor: Critical Insights*, ed. E. Osei- Prempeh E. Osei Kwadwo, Joseph Mensah, and Senyo BS. K. Adjibolosoo, 121–142. Aldershot, Hants, UK; and Burlington, VT: Ashgate.

———. 2008. Media Pluralism, Democratic Discourses and Political Accountability in Africa. Paper Presented at the Harvard Workshop on the Role of the News Media in the Governance Agenda. http://ksghome.harvard.edu/~pnorris/Conference/Conference%20papers/Tettey%20Africa.pdf.

Tettey, W. J. and K. P. Puplampu. 2005. “Border Crossings and Home-Diaspora Linkages among African-Canadians.” In *The African Diaspora in Canada: Negotiating Identity and Belonging*, ed. W. J. Tettey and K. P. Puplampu, 149–174. Calgary: University of Calgary Press.

Tsaliki, L. 2003. “Globalization and Hybridity: The Construction of Greekness on the Internet.” In *The Media of Diaspora*, ed. K. H. Karim, 162–176. Oxford, UK: Routledge.

Walmsley, D. 2000. “Community, Place and Cyberspace.” *Australian Geographer* 31(1): 5–19.

CHAPTER 10

Globalization from below? ICTs and Democratic Development in the Project "Indymedia Africa"

Fabian Frenzel and Sian Sullivan

Indymedia Africa (IMCA) is a global network of media activists that aims to both connect and foster the use of Independent Media in Africa. Originating in the digital age activism of the late nineties, the Indymedia network has been surfing a wave of optimism regarding the potentials of new media and the digital public sphere to democratize publishing and the media. Information and Communication Technologies (ICTs) were understood as permitting "convergence" between people and movements in a horizontally organized fashion, thereby facilitating desired organizational cultures based on consensus and plurality, and producing "open spaces" relatively unstructured and uncontrolled by conventional political and economic structures. As an element of a "globalization from below," IMCA considered these ideas as an answer to problems of democracy and freedom of expression in Africa and attempted to spread its own organizational principles into African independent media. In four years of creating virtual and physical convergence spaces, online forums, and Web sites, as well as organizing transnational gatherings, however, the IMCA network has had to face something of a reality check regarding the conditions of its own work and the African context. It has also gone through a process of action and reflection that appears symptomatic for a variety of initiatives of global cooperation in the field of new media, highlighting the limits of technological and pragmatic answers to the debate of the democratic potentials of these media. This chapter considers the actors and ideologies that have informed, defined, and altered IMCA practices since its inception, via an action research guided analysis of its virtual and physical encounters.

Introduction

Independent media projects such as the global Independent Media (or "Indymedia") Network (www.indymedia.org), which are based on the use of new ICTs, have been part and parcel of the transformation of publishing in the global sphere in the digital age. Born as a tool of communication between activists in the protests against the World Trade Organization in Seattle in 1999, and following prototypes such as the use of a shared Web site for uploading media reports during the London protests against the G8 on June 18, 1999 (Notes from Nowhere 2003, 231–2), Indymedia attempts to be an "open space" in the virtual world (Pickard 2006; Mamadouh 2004; Pickerill 2004). "Open space" here refers to the intention that Indymedia functions as a news-based Internet portal that allows the "open posting" of articles by any author to a Web site, while a set of publishing regulations and an inclusive editorial collective are the only gatekeepers (Keraghel and Sen 2004; Patomäki and Teivainen 2004; Böhm et al. 2005). Instead of depending on corporate or state sponsored media, grassroots initiatives and individual activists thus find here an open media space where their localities and concerns can be independently produced and represented for a potentially global audience, facilitating the networking of similar and related struggles, and allowing for a radical horizontal reorganizing of the public sphere. The open editorial collectives operate through online synchronous communication (Internet Relay Chats or IRCs) to allow participants to engage in the editorial process and enable consensus resolutions to be reached. Despite the importance of ICTs in the communication structures of Indymedia, its principles also emphasize a strong local and face-to-face component of the editorial work. Essentially a network of autonomous local groups, Indymedia as an organizational "umbrella" insists on a certain purity with regard to new members who want to join the network.[1] Indymedia also attempts to foster "globalization from below," a concept that derives from a critique of a top-down neoliberal globalization driven by the globalized finance sector, large-scale corporations and powerful states (Della Porta 2006; Harvey 2007; Klein 2007).

Technically, Indymedia's development was permitted and propelled by the rapid expansion of the availability and use of ICTs. In fact, the ever-increasing number of people using ICTs at decreasing costs constitutes one of the major reasons social theorists speak of systemic changes in social organization under current rapid processes of globalization (e.g., McLuhan 1964; Dery 1996; Melucci 1996; Castells 1997; Gleick

1999). The ability to use alternative and new media thus is critical in producing emerging forms of political organization, particularly in the realm of contemporary social movements (Escobar 2004, 4). Indymedia is a prime example of this kind of "information-age activism" (Routledge 2003). What is often overlooked in this celebration of ICTs, however, is the fact that they have been equally crucial to the development of neoliberal globalization, particularly in the realm of international finance. ICTs are integral not just for democratic critique, but also for the development of a late capitalist global economy. The centrality of ICTs in both globalization and its critical discourse opens a set of questions regarding a situation of limited access and spread of ICTs, as is the case in Africa.

Development, "Neoliberal" or "from below"?

Although the growth rate of Internet usage and available bandwidth are massive today in Africa as elsewhere, there is a dramatically smaller availability and use of the Internet in Africa than in any other part of the world.[2] Globally, there is a digital divide,[3] or inequality, in Internet usage, related to a lack of technostructure.[4] This has been identified as a major development concern (Flatz 1999) resulting in such techno-deterministic projects as Negroponte's "Laptop for every child" some of which are highly uncritical of the ideologies underlying media, education, and technology programs in development projects. (Berman 1980a, 1980b). This is particularly in the area of producing a hegemonic global civil society shaped by Northern discourses of trade liberalization, individualistic consumerism, and "democracy" (cf. Reed 1996; Gramsci 2003 [1971]; Spicer et al. 2006).

The global state-critical and neoliberal discourse resulted in a tendency to bypass state institutions in international funding in the 1990s, and to rely more on civil society actors such as NGOs. Non-state actors were believed to be less biased, less bureaucratic, less corrupt, and more reliable partners in aid than state agencies. But the new focus on civil society did little more than legitimate structural adjustment along the lines of the Washington Consensus. State spending in the South was reduced or terminated, education, health, transport, and the environment, were privatized, and nonprofitable aspects of the state's solidarity and transfer systems were reorganized through NGOs that became the main receivers of international aid flows. These new NGOs were often the antennae of transnational, Northern-dominated NGOs. Wallerstein (2004, 269) thus points out that despite the anti-systemic, or state-critical origin of several NGOs in the North, their actual policy and impact in their international work have made them appear more as "agents of their home states" in the South. They were certainly agents of the neoliberal turn in global aid flows. This has not necessarily delivered a higher rate of development, democracy, or alleviation of poverty for Africa. From the perspective of producing greater social justice in Africa, the neoliberal development policy arguably has failed dramatically (Shikwati 2005; Mwenda 2006).

Indymedia is centrally engaged in providing, supporting, and disseminating the critique of neoliberal globalization; however, in its own network it also is confronted with a "development" problem. There is a digital gap mirroring the global one in the Independent Media network, which currently has some sixty-one Independent Media Centers (IMCs) listed for the United States, fifty for Europe (with a further fourteen listed for the UK), but only six for Africa.[5]

While critical of neoliberal policies and resulting development strategies, Indymedia also found itself creating its own "development" initiative. The IMCA network is animated by a wish to contest and depart from criticized neoliberal development strategies and to create an alternative approach. IMCA thus pursues what has been described as a need to go beyond development (Escobar 1992), particularly by focusing on grassroots initiatives as central agencies in the challenge to "co-move" rather than to develop (Esteva 1987, 33). The idea of "co-movement" is to "intensify the processes of construction of direct democracy." IMCA attempted to put into practice these ideas in its four years of existence.

IMCA

Founded in 2003, the IMCA working group consisted of African and European members of the global Indymedia network who shared the aim of holding a conference of Independent Media Activists in Dakar, Senegal. IMCA started from the premise that it would be a good idea for existing African independent media initiatives to be linked with Indymedia, while at the same time hoping to foster the creation of new initiatives.

In the making of the first conference, it became clear that the attempt to depart from the problems associated with neoliberal development policies had failed in several aspects:

1. In Dakar local participants were not involved in the planning process. The whole idea of the conference was conceived outside Dakar. Larger local participation only commenced with the beginning of the conference, thanks to an outreach initiative to local students and some local activists. The *Global-Local Nexus* or the politics of scale in transnational planning process was radically imbalanced;
2. As the money for the meeting came from the "north," it was mainly Northern participants who controlled and administered

the spending of funds and acted as *organizers* at the convergence space (see below). Additionally, Northern participants knowledgeable in ICTs were invited to join the meeting, specifically coming to *teach* participating Africans. This resulted in a situation where the power lines at the convergence space were along a North-South trajectory.

3. Several Northerners participated in the meeting on their own terms, using their own resources. No nonlocal Africans were able to participate as independently as this. Instead, their participation was enabled by funding from project budgets. Radically different *global mobilities* structured the conferences and influenced actor's attitudes and roles in it.

Since the Dakar meeting, IMCA has organized two similar conferences: the IMC at the polycentric World Social Forum (WSF) in Bamako, Mali, in January 2006; and at the WSF in Nairobi, in January 2007. Through taking a closer look at the 2007 Nairobi conference, we will discuss the process and the learning experience of the IMCA over the period of its existence, with reference to these three points.

Analyzing Convergence Spaces: WSF Nairobi 2007—an Ethnographic Exploration

The concept of "convergence space"—used broadly in globalization critical movements to describe their open-space meetings and gatherings—has been proposed as a "tool by which to understand and criticize grassroots globalization networks" (Routledge 2003, 334). Amongst activists, the phrase refers to specific locations—sleeping spaces, media spaces, and so on—at the large-scale gatherings of protestors occurring at summit mobilizations such as the anti-G8 protests. The term reflects an understanding that in these meetings activists and groups from diverse backgrounds and cultures come together to form a new social entity, respecting the initial differences rather than subsuming them into one overall narrative. At the same time, these real-time meetings are preceded, prepared, enhanced, reflected upon, and ultimately made possible by ICTs situated in what has been labeled convergence culture (Jenkins 2006). Mailing lists, web chat-rooms and wiki-pages, open publishing, and web 2.0 allow the formation of global networks that are able to mobilize people for real-life encounters.

The following case-study focuses on the convergence space of the Nairobi IMCA conference, an actual meeting of not more than forty-five people at a time that took place over three weeks in Nairobi, as well as on the virtual social spaces that existed around this convergence and that made the actual meeting possible. As a place of negotiation of heterogeneity and the politics of scale, the conference and its virtual environment highlights some of the issues produced more generally by emerging global interactions and transfers of aid, technology, and discursive frames. As Routledge (2003, 347) points out, "attention needs to be paid to the internal structures of the movements and groups that participate in convergence spaces, and to their placing within local realities." In this respect the notion of "open space" is significant in the analysis of convergence spaces. The particular role that the ICT-based virtual convergence plays in relation to the physical convergence is discussed here in some detail.

The IMCA embraces the need to reflect on its own dynamics, and participants in the Nairobi convergence gave their consensus to this study, which thereby owes many of its insights to the ensuing collective reflections. We approached the convergence as participants and observers and with "action research"-guided qualitative methods,[6] applied discourse analytical tools to public mailing lists and online forums (e.g., Fowler 1991; Fairclough 1995), and explored peoples' motivations and perspectives in a set of interviews with a total of ten participants.[7] In the following sections, only a very small amount of the material is displayed. The quoted sources were chosen because they reflected best the tendencies and arguments that were exchanged in the convergence space. The choice of the largely ethnographic and qualitative methods was necessitated by the impossibility of quantifying the largely hidden power structures and contradictions described above. Insider knowledge and a long-term commitment to the work of the group helped to understand the specific meanings carried in expressions and discourse.

Our analysis is oriented toward the three points identified above. In each case a discussion follows the ways the specific problem occurred, and how it influenced and prevented, helped or necessitated, the ideal of a horizontally and democratically organized convergence in the case of the Nairobi meeting. By drawing on participant observation and discursive analysis of the Dakar conference and the Nairobi event, their planning and realization, and their material and virtual manifestations are explored. The changing role of ICTs in this process is of specific importance. The collective learning process of the group over a period of several years is of additional interest here. In particular, we highlight some ways in which actors might be drawn into roles that can reinforce rather than reconfigure the material realities of global inequality, despite their awareness of these patterns and processes. The reflections in this study thus seek to contribute to the ongoing organizing work with IMCA.[8]

Global-Local Nexus

From the moment the IMCA was formed at the convergence spaces of the polycentric WSF in Bamako in 2006,[9] people within and beyond the

IMCA working group discussed plans to create a larger and well-planned convergence-based IMC at the WSF in Nairobi. The actual discussions and planning process started in the early summer of 2006 through the connection of the already established IMCA working group with the Philadelphia, U.S.-based community radio project, Prometheus. Other than in Bamako, the IMCA working group was already networked into Kenya through the Kenya IMC, which had been established as a result of the Dakar Indymedia conference in 2004.[10] From September 2006, a call out[11] was circulated within and beyond the Indymedia network to join the convergence in Nairobi. Funding applications were made while the planning process continued through communications on the mailing list and wiki of the IMCA working group. The overall plan was to connect as many media activists as possible globally, and especially to assist with the participation of African media activists through the network and the funding possibilities it might tap into. The specific importance of actual meetings to build transnational networks has been stressed above. The global face-to-face convergence of media activists from different cultures and backgrounds was framed as part of the larger convergence of "insurgent cosmopolitanism" in the World Social Forum (de Sousa Santos 2006).

As for the local impact, it was hoped that the convergence would boost the IMC Kenya group, dormant since January 2006, and allow the Kenya IMC to network further with local media activists. The IMCA was also hoping to enlarge and strengthen its own network by allowing other African delegates to participate, especially from the long-running IMCs in South Africa and Ambazonia/Cameroon, as well as the new IMCs in Mali and Nigeria. The overall aim was to create an outstanding experience for everybody involved: an experience of a horizontal organizing process as much as of active participation in the WSF. Practically a series of discussions and seminars were planned on various media-related issues, while hands-on workshops were to teach radio construction, silkscreen printing, open publishing, and video editing.

The planning process was largely ICT-based and transnational; however, following the experiences in Dakar, it was clear that a strong local planning in Nairobi was needed just as much. Reflecting on the planning process in Nairobi, one of the main local organizers thus remembered that

> We could make a difference and it was up to us in Kenya to ensure that we actually provided leadership to make it happen (Interview 9).

However, the process appeared to be problematic. The local preparation fell to two main organizers. Of these one increasingly was unable to provide any support, because he was involved in the WSF organizing himself, while the other was committed to a full-time job up until ten days before the convergence started. Other local volunteers were not ready to take the same amount of responsibility and leadership in the organizing process

> which was actually a source of frustration because as I said we knew we had to provide leadership so here was a case where people were enthusiastic and sometimes it wasn't always obvious to them that there were so many other things we were doing here (Interview 9).

IMC Kenya was expected to be not only the main local logistical organizer but also was organizing the content of the event. This was particularly in relation to preparing the discussion and discourse of the conference, the latter aspect central in contributing to the philosophy of a "globalization from below." These tasks overwhelmed the local structure due to the labor and time constraints mentioned above.

This situation caused several problems for the project. Because certain logistical aspects were not dealt with, the project lost resources. One example of this was the process of obtaining visas for delegates from West Africa. This task was not dealt with in time and resulted in the inability for these delegates to attend at all. As tickets for them were purchased in the belief that the visa issue was being dealt with, the project lost a significant amount of money in cancellation fees, and also in time spent dealing with the issue. But more importantly, the participation of five West African delegates became impossible, causing a major setback for the within-Africa networking process, and thereby seriously hampering one of the central aims of the convergence.

In other words, the local group was not strong enough to support the logistical necessities of a conference of this size. This problem illustrates how material conditions structure local organizing culture, as well as how the celebration of transnational and somewhat delocalized ICT-based organization can foster a gross misinterpretation of logistical problems on the ground. Global ICT-based communication possibilities seem to provide a background for common and horizontal action on the ground, yet the particularities of localities may easily be ignored, masking and maintaining strong structural imbalances with regard to power and resources in the wake of a global organizing process. The colonizing notion of "open space" (Chesher 1999) seems to become obvious here, as the global ICT-based communication gives an illusion of open participation in a planning process that actually overwhelms the local group. One option might have been to strengthen the local organization by paying full-time local organizers. This was not seriously considered, as Indymedia does not—in general—employ professional staff. In any case, this professionalization would have amounted to the reestablishing of postcolonial structures in the African context, as local network nodes would have become a contractor of the globally sponsored Indymedia network, thereby contradicting the idea of a "globalization from below."

Instead, a second option came into play, with local organizing in the case of the IMCA conference becoming dependent on the help and assistance of Northern volunteers. The value of this assistance—being nonlocal and not speaking the local languages—was limited in practical terms. Interviewee 9 did comment on the steep but valuable learning experience that was fostered through having to respond directly and urgently to the unexpected logistical issues arising as the project unfolded, understanding the "madness and craziness" of the IMCA convergence space as a chance for Northern participants to experience better the real local issues at stake, or what was described as "the African context" by many of them. This somewhat conciliatory and optimistic view, however, contrasted sharply with that of another Southern participant, who described the convergence space as a situation of "constant crisis," and argued that a productive negotiation of difference was unable to take place as a result.[12] He concluded that the North-South divide was reinforced as a power-divide precisely because there were such massive logistical challenges, which thereby allowed Northern participants to assert expert roles as "technocrats" for pragmatic reasons. This observation takes us neatly into our next point.

Organizers and Teachers

The imbalance of organizing responsibilities and power along the lines of existing global structures was matched by a stream of reflections, critique, and explanations amongst a group clearly motivated by global justice issues. A Southern participant critiqued the situation, in retrospective, vociferously and provocatively, and his critique was representative of many of the statements people gave about the encounter:

> from start to finish, the convergence lived in a situation of crisis...an environment in which experts and bureaucrats flourish their capabilities, and a particular kind of emergency community thrives while democracy is postponed, an emergency community in which those who typically have more power—due to race, class, a command of imperial languages—consolidate that power...it's not surprise that this technocracy was white.[13]

In the convergence, an early attempt was made to put structures in place to deal with key logistics democratically and to assess and discuss the varying expectations and roles of participants to allow a collective reflective process regarding this. Daily plenary meetings and several working groups for various logistical issues were put into place. Additionally, sessions were devoted to debating expectations and objectives of the convergence. Importantly, for all aspects of the logistical and content side of the convergence, structures had been formed in the course of the first week of the meeting, while the daily plenary session remained to decide substantial issues, by consensus where possible.

These structures are based on the organizational patterns of Indymedia as it was developed in a historical process as part of the consolidation of a Northern activist culture in recent years. From this it seems they can hardly be separated: the processes of decision-making seem not so much to depend upon the nominal structures and the ideological claims of their horizontality, but on experience, individual performance, and a practical situation that unfolds. Interviewee 9 addressed the issue of control of financial resources in the convergence. Despite horizontal ideals and structures, Northerners technically held the funding and thus were perceived by Southerners as gatekeepers; as controller of funds rather than as equal participants in an open debate about their use (Interviewee 9). Furthermore, the use of most of the funds was not controlled by the actual convergence on the ground. but by the ICT-based planning process in advance of the actual meeting. Although this was done democratically, Southern participants were underrepresented in the planning process due to their limited access to ICTs. In this way, the ICT-based preparation made a democratic approach to the physical convergence difficult.

In summary, Northerners holding funds were perceived by some Southerners as unwilling to share responsibility for funds democratically. As such, Northerners were perceived as obstacles to the emergence of a properly democratic and equitable process (Interviewee 9). An additional context was the case of a set of participants from Uganda, who actively tried to bargain for personal allowances out of the collective resources in nearly every collective meeting. Within the group, this undermined trust and beliefs in the seriousness of their alternative media concerns.

Many Northerners also received requests for material aid outside the official structure of the convergence. A U.S. media activist (Interview 5) described how a roommate "asked me constantly if I could give her everything [electronics] I had," without a clear idea of what to do with them, especially the ICT gadgets. This interviewee was concerned about articulating a distinction between empowerment and charity, which was expressed in comparable terms by several Northern participants. While she felt unsure about the need and use of charity, as the practice of materially enriching people for their benefit, she found empowerment of people through skills training to be important. The training she provided the African participants in interview skills was something that she considered desirable, as it could help them enrich their lives (Interview 5).

At the same time, several of the Southern activists expressed in interviews that they were expecting help and support from the Northern participants in the use of technologies. As stated by a Nairobi-based participant, his desire was to connect with "professionals from all over the world" who can "give us the knowledge" in a situation where "we don't have the resources and facilities." He also suggested that he understood the encounter to be an exchange process of global skills for local experience, in which the teachers "can learn more of our local things," thereby referring to sites of tourist interest for Northerners, such as the Kenyan national parks (Interview 2). Another African participant explained that training in basic radio or interview skills, as it was done in the convergence, offered a great opportunity for them to improve their knowledge, and emphasized the monetary value of the teaching. In their environment, similar training was available only for high payment (Interview 6).

As such, ICTs became something like alternative currencies that structured the convergence. They were techniques of empowerment that facilitated a transfer of resources in ways that all those involved felt more comfortable with than if this had come down to simply handing over material and money. This transfer worked for "both sides," and it came to be prioritized over the ideals of horizontality and "open space." The U.S. media activists gave a pragmatic notion of this point:

> The situation is when you have a group of people that come in and they have got access to not only the equipment but basic skills around media production. And there are a group of people who are really hungry for those skills and they are vocal about wanting those skills now. Do you spend a whole bunch of time trying to work out horizontal leadership, oh I am not really a teacher we are all teachers. Or do you pull up your sleeves and say lets learn audio production today (Interview 5).

The role of ICTs in this context, however, is very different from the "information age activism" vision discussed at the start of this chapter. They did not democratize the convergence as such, and thereby radically reorganize the public sphere. Rather, they were unable to escape the structuring effects of preexisting inequalities between Northern and Southern activists. As such, it could be argued that ICTs in this context became a currency that facilitated the exchange between people in a situation that was dominated by large inequality. ICTs thus became objects of a unidirectional flow of nonmonetary resources channeled from North to South: they became parcels of aid, with Northern participants, as organizers and teachers, ultimately deciding what constituted charity and what might facilitate empowerment, or what was acceptable aid and what was not.

As we saw, the contradictions of aid in relation to the ideals of the project were discussed intensely in the context of organizing and teaching the convergence, and vociferous critique was voiced in regard to the emergence of postcolonial patterns of North-South relations. To discuss this further, we now consider a third area of contradiction, that of differential global mobilities. In Nairobi, the group seemed little concerned about the practices of sponsored mobility that allowed many of the participants to actually meet, despite the fact that this aid was also clearly structured along the lines of the North-South trajectory.

Global Mobilities

Indymedia's principles of open participation and open space are generally applied to both virtual and physical meetings. In the case of physical meetings, nonlocal participants need to travel to arrive. Developed in the North against a backdrop of a large volunteer force of political activists that are interested in participating in open conferences and are willing and able to take care of most of the costs involved, these principles are discriminating against people whose access to mobility is financially constrained. Indymedia has provided funds in a variety of forms to allow people to be supported in attending meetings across different geographical areas within the Global North and beyond, but these systems are marginal in comparison to a much larger number of self-funded participants who take part in meetings. In the African context, however, the number of people ready to spend time and money to participate in social movements is much smaller than in the Northern areas where Indymedia originated. Acknowledging these problems, but also unwilling to create a clear distinction between Northern and Southern participants, IMCA therefore attempted to fund the travel of all participants for its initial conference in Dakar. Practically, however, it proved to be much easier to secure funding for African media activists. The reason for this was that the German foundation Umverteilen—one of the most important funding sources for all three convergences—has a policy to support only African participants.

This system of positive discrimination had been established by the time of the Nairobi conference as the standard procedure for the mobility policy in IMCA. The general budget of the convergence covered all travel and other costs for African participants. The fact that the project's policy on travel support is formed by its main funding source shows again how preexisting structures come to govern and structure new projects in the field. At the same time, it was critical in terms of enhancing African participation. In the case of the Nairobi meeting, it meant that the participation of Kenyan and other African delegates in the convergence was higher than that of the Northerners (24 to 18), and this also accounted for the largest part of the funding, consuming about one-third of the overall budget.

Interestingly, after some debates regarding this issue in Dakar,[14] positive discrimination of African delegates in regard to mobility hardly elicited any critical discussion during the Nairobi convergence. However, it constituted a clear dividing line between Southern and Northern participants and a one-directional flow of aid, as none of latter benefited from the IMCA budget in a similar way.

As in Dakar, the participation of Northern activists in Nairobi was partly voluntary and self-organized. While most of those coming to fulfill explicit trainer roles were funded, this was not done through the overall budget of the conference, but through funds that these participants had managed to organize for themselves. Additionally, there were other Northern participants who came solely on their own behalf, without any funding. All participants benefited from free food and housing in the convergence, but they also shouldered private travel costs and other financial burdens that were incurred while attending. In part, this is because financially they were able to do so; and in part this also speaks to participants' desires to proliferate independent media content and organizations in the "global south." Another view might also place such participations in the realm of tourism, or, more precisely, of "voluntourism," which Wearing (2001) defines as global volunteering, involving high mobilities. Volunteer mobility has been identified as important for social movements in enabling the growth of social networks and the sharing of information regarding issues of concern (McGehee and Santos 2005).

In the convergence, "tourist" and "tourism" became terms to negotiate the seriousness of participants' concerns. Some participants were accused of acting as if on "a European funded excursion." Another participant proposed a typology of activists in this context. According to him, political events tend to consist of "real activists" actively organizing political events, in contrast to the "political tourists" who were just participating (Interview 1). It becomes obvious that the role of a "mere" tourist is something that participants might want to distance themselves from in these kinds of meetings. Conversely, being beyond the label of "tourist" might mean to render the high mobility employed to allow for participation in meetings as somehow purposeful and functional for the causes of the global justice movement.

As with the role of ICTs discussed in the previous section, mobilities employed in the convergence thus were at a crossroads between a technique to facilitate a "globalization from below" on one hand, and acting as a parcel of aid, on the other. To some extent they affirmed structures of global inequality, reinforcing a dividing line between Northerners and Southern participants. At the same time, no one questioned that mobility needed to be supplied to African participants through a system of positive discrimination. Instead, participants discussed the way mobilities were actually used, whether they were used for mere tourism or for the purpose of active involvement in the convergence and therefore ultimately for the advancement of global justice. Importantly, this debate on the purposefulness of the employed mobility, whether funded or self-arranged, was applied equally to all participants regardless of their backgrounds. It could be interpreted that the project IMCA matured over the initial problems with positive discrimination in the context of mobility and discussed this matter more thoroughly than the question of the transfer of ICTs and related skills. It may be that this reflects the specific ideological configuration of ICTs in the context of the IMCA project, as well as the broader contexts that structure ICTs as quick answers to questions of global inequality and development.

Discussion

Experience and critique of international aid projects and initiatives indicate that these initiatives may enhance the inequalities they are intended to ameliorate. Despite a high critical awareness of these problems of aid and the conscious attempt to "develop from below," the IMCA also reproduced global inequalities in the myriad articulations between "North" and "South" via existing structures of inequality that permit Northern activists to transmit certain ideas and practices, even as Southern participants also shape these through their engagement.

Potentially then, Indymedia in Africa, as a political project in the realm of ICTs and new media, may underestimate the significance of structural inequalities in its own network and practices. As this case suggests, postcolonial and neoliberal realities to some extent were perhaps enhanced by idealistic assumptions regarding the potential of ICTs to overcome these realities in what is framed as "open space" and "information-age activism." ICTs to some extent become a currency in which global inequalities are traded and maintained, rather than tools to overcome them. At the same time, organizational principles of "open space" and "volunteering," as developed in the North, are based on a priori similarities in backgrounds and a shared embeddedness in relation to a corresponding technostructure; they are culturally and socially specific manifestations that are fairly limited even within Northern societies. When they are uncritically promoted elsewhere, they can take place in the realm of what has been described as a "double colonization": the way the project is funded already operates in the context of aid flows and associated controversial practices, while the idealisms linked with uncritical assumptions of ICTs as a means to overcome these structures might render practitioners less conscious of the persistence of postcolonial structures in the project.

ICTs specifically, and technology in general, thus may shift from being a potential solution to becoming a problem, particularly when their absence becomes the explanation for deficits in democracy or development that might be more likely explained by structural conditions under global capitalism or problematic leadership issues. While ICTs might empower grassroots groups, networks, and individuals, their availability and proficiency does not necessarily result in empowerment. In the case of ICTs in Africa then, their potential is limited by the same constraints that caused their delayed arrival in the first place.

It is interesting to see how the convergence dealt with the imbalance in access to mobilities that allowed for people to actually meet in Nairobi. The clear positive discrimination of African participants was not questioned, for it was obvious that, compared to Indymedia activists in the North, constrained resources prevented Africans from participating outside their own countries. Instead, the crucial debates centered around questions of how the mobilities were actually used. Did the enhanced mobilities of the African participants and the volunteer mobilities of Northern participants facilitate what might be considered as "political tourism"—a privilege that is fully part and parcel of the global economic *status quo*; or did they enable contexts of empowerment that might flow into a "counter-globalization from below"?

In face of the differences that exist in access and knowledge to ICTs in different parts of the world, positive discrimination to balance these inequalities is a path toward enabling social convergences and the strengthening of critical discourses and practices of empowerment. Critical questions for IMCA might be the following: How are the transfers used? Did the participants embed the enhanced IT knowledge into local practice? Did ICTs in a given situation constitute mere objects of status, reproducing global inequalities? Or did they enable counter practices and organizations of globalization? By asking such questions, projects such as the IMCA become part of a process that moves towards contesting and remolding the structural inequalities in which it is itself embedded.

Notes

1. Indymedia emphasizes the importance of horizontality for its working structure. Collectives and local IMCs that want to become a node in the network have to adhere to these and a few other principles. For details see https://docs.indymedia.org/view/Global/NewIMCForm (accessed June 1, 2008).
2. Although differences between countries are significant, most countries in Africa seem to face similar problems. Mike Jensen's webpage gives a comprehensive insight into the development of Internet connectivity

in all parts of Africa, documenting impressive growth but also a continuing lack of resources, http://www3.sn.apc.org/africa/ (accessed June 1, 2008).

3. See http://en.wikipedia.org/wiki/Global_digital_divide (accessed June 1, 2008).
4. East Africa, for example, currently relies on satellite communications, and is awaiting the completion of the construction of a fibre-optic submarine cable. As stated on the Cipesa Web site (2006, 2), which documents the efforts and aims to increase the ability of Africa and Africans to utilize new ICTs, "[t]his situation made East Africa one of the most 'digitally excluded' regions of the world (. . .)," see http://www.cipesa.org/195 (accessed June 1, 2008). Further delays of the construction of the fibre-optic cable were recently reported, http://news.bbc.co.uk/2/hi/africa/5328578.stm (accessed June 1, 2008).
5. See www.indymedia.org (accessed June 1, 2008).
6. For example, see action research resources at http://carbon.cudenver.edu/~mryder/itc/act_res.html (accessed June 1, 2008).
7. These interviews were conducted during the course of, or immediately after, the convergence in Nairobi. All participants are kept anonymous here.
8. A detailed evaluation of this process is also available online at https://en.wiki.in-no.org/WSF2007Report (accessed June 1, 2008).
9. See http://en.wiki.in-no.org/WsfMali (accessed June 1, 2008).
10. See http://www.kenya.indymedia.org/ (accessed June 1, 2008).
11. See http://lists.indymedia.org/pipermail/imc-africa/2006-September/0911-qt.html (accessed June 1, 2008).
12. In an e-mail, see http://lists.indymedia.org/pipermail/imc-africa/2007-February/0204-mx.html (accessed June 1, 2008).
13. See http://lists.indymedia.org/pipermail/imc-africa/2007-February/0204-mx.html (accessed June 1, 2008).
14. In http://archives.lists.indymedia.org/imc-africa/2004-March/000275.html (accessed June 1, 2008).

References

Adero, B. 2007. "World Social Forum Delegates Expected to Strain Host City." *The East African Standard,* January 1, Nairobi section. Nairobi: Standard Media Group.

Appadurai, A. 2000. "Grassroots Globalization and the Research Imagination." *Public Culture, Special Issue on Globalization* 12(1): 1–19.

Ayittery, G. 2002. "Why Africa is Poor." *Daily Telegraph*, August 26.

Berman, E. H. 1980a. "Educational Colonialism in Africa: The Role of American Foundations, 1910–1945." In *Philantropy and Cultural Imperialism. The Foundations Home and Abroad*, ed. R. F. Arnove, 89–110. Boston, MA: G. K. Hall & Co.

———. 1980b. "The Foundations Role in American Foreign Policy: The Case of Africa, Post 1945." In *Philantropy and Cultural Imperialism. The Foundations Home and Abroad*, ed. R. F. Arnove, 111–135. Boston, MA: G. K. Hall & Co.

Böhm, S., S. Sullivan, and O. Reyes. 2005. "The Organization and Politics of Social Forums." *ephemera: theory and practice in organization* 5(2). http://www.ephemeraweb.org/journal/index.htm (accessed June 1, 2008).

Brown, S. and X. Lehto. 2005. "Traveling with a Purpose: Understanding the Motives and Benefits of Volunteer Vacationers." *Current Issues of Tourism Concern* 8(6): 479–496.

Castells, M. 1997. *The Power of Identity*. Oxford: Blackwell.

Chesher, O. 1999. "Colonizing Virtual Reality: Construction of the Discourse of Virtual Reality, 1984–1992." *Cultronix*. http://cultronix.eserver.org/chesher (accessed June 1, 2008).

Della Porta, D., M. Andretta, L. Mosca, and H. Reiter. 2006. *Globalization from below Transnational Activists and Protest Networks*. Minneapolis: University of Minnesota Press.

Dery, M. 1996. *Escape Velocity: Cyberculture at the End of the Century*. London: Hodder & Stoughton.

Escobar, A. 1992. "Imagining a Post-Development Era? Critical Thought, Development and Social Movements." *In Third World and Post-Colonial Issues*, 20–56. *Social Text*, 31–32. Madison, WI: Coda Press.

Escobar, A. 2004. "Beyond the Third World: Imperial Globality, Global Coloniality and Anti-Globalization Social Movements." *Third World Quarterly* 25(1): 207–230.

Esteva, G. 1987. "Regenerating People's Space." *Alternatives* 12(1): 125–152.

Fairclough, N. 1995. *Critical Discourse Analysis*. New York: Longmann.

Ferguson, J. 1994. *The Anti-Politics Machine: 'Development,' Depolitization, and Bureaucratic Power in Lesotho*. Minnesota: University of Minnesota Press.

Flatz, C. 1999. Internet in Africa—Africa *auf der Ueberholspur der Entwicklungsautobahn*? *Interasia*.http://www.interasia.org/vietnam/hanoi_afrika.html (accessed June 1, 2008).

Fowler, R. 1991. *Language in the News: Discourse and Ideology in the Press*. London: Routledge.

Gleick, J. 1999. *Faster: The Acceleration of Just About Everything*. London: Vintage Books.

Gramsci, A. 2003 (1971). *Selections from the Prison Notebooks of Antonio Gramsci*. London: G. Lawrence and Wishart.

Harvey, D. 2007. *A Short History of Neoliberalism*. Oxford: Oxford University Press.

Jenkins, H. 2006. *Convergence Culture: Where Old and New Media Collide*. New York: New York University Press.

Keraghel, C. and J. Sen. 2004. "Explorations in Open Space: The World Social Forum and Cultures of Politics." *International Social Science Journal Special Issue* 182.

Klein, N. 2007. *The Shock Doctrine. The Rise of Disaster Capitalism.* London: Metropolitan Books.

Mamadouh, V. 2004. "Internet, Scale and the Global Grassroots: Geographies of the Indymedia Network of Independent Media Centers." *Tijdschrift voor Economische en Sociale Geografie* 95(5): 482–497.

Marquez, H. 2006. "Series of Global Protests to Begin in March." *Peace No War.* http://peacenowar.net/newpeace/index.php?option=com_content&task=view&id=80&Itemid=0 (accessed June 1, 2008).

McGehee, N. G. and C. A. Santos. 2005. "Social Change, Discourse and Volunteer Tourism." *Annals of Tourism Research* 32(3): 760–787.

McLuhan, M. 1964. *Understanding Media: The Extensions of Man.* Corte Madera: Gingko Press.

Melucci, A. 1996. *Challenging Codes.* London: Cambridge University Press.

Mwenda, A. 2006. "Foreign Aid and the Weakening of Democratic Accountability in Uganda." *Foreign Policy Briefing* 82. http://www.cato.org/pub_display.php?pub_id=6463 (accessed June 1, 2008).

Notes from Nowhere Collective, eds. 2003. *We are Everywhere: The Irresistible Rise of Global Anti-Capitalism.* London: Verso.

Patomäki, H. and T. Teivainen. 2004. "The World Social Forum: An Open Space or a Movement of Movements?" *Theory, Culture and Society* 21(6): 145–154.

Pickard, V. 2006. "United Yet Autonomous: Indymedia and the Struggle to Sustain a Radical Democratic Network in Media." *Culture & Society* 28: 315–346.

Pickerill, J. 2004. "Rethinking Political Participation: Experiments in Internet Activism in Australia and Britain." In *Electronic Democracy Mobilization, organization and participation via new ICTs*, ed. R. Gibson, A. Roemmele, and S. Ward, 99–129. London: Routledge.

Reed, D., ed. 1996. *Structural Adjustment, the Environment, and Sustainable Development.* London: Earthscan Publications.

Routledge, P. 2003. "Convergence Space: Process Geographies of Grassroots Globalization Networks." *Transactions of the Institute of British Geographers* 28(3): 333–349.

Shikwati, J. 2005. "For God's Sake, Please Stop the Aid." *Spiegel Online* (July 4). http://www.spiegel.de/international/spiegel/0,1518,363663,00.html (accessed June 1, 2008).

de Sousa Santos, B. 2006. "Globalizations." *Theory, Culture & Society* 23: 393–415.

Spicer, A., S. Sullivan, and S. Böhm. 2006. Becoming Global (Un)Civil society: The Case of Indymedia. Paper Presented at the Internet Institute, March 2006, Oxford University, UK.

Wallerstein, I. 2004. "New Revolts Against the System." In *A Movement of Movements—Is Another World Really Possible?*, ed. T. Mertez. 215–245. London: Verso.

Wearing, S. 2001. *Volunteer Tourism. Experiences that Make a Difference.* Oxon: Cabi Publishing.

CHAPTER 11

New Public Spheres: The Digital Age and *Big Brother*

Keyan G. Tomaselli and Ruth E. Teer-Tomaselli

"Reality Television," a novelty only a few years ago, is now an entrenched genre, and has been afforded the same status as long-established formats such as soap operas and sitcoms (cf. Hill 2002; Mathijs and Jones 2004; Tincknell and Raghuram 2002). Our own study attempts to examine the way in which ordinary people, politicians, civil society, and religious organizations have responded to these programs.

Reality Television is unashamedly voyeuristic, sensationalist, and contrived. The genre purports to portray the illusion of unmediated events in which contestants compete for the public vote. Particularly in its earlier manifestations, salaciousness was a built-in characteristic. Jeremy Daniel, in "Being Big Brother," writes without a trace of irony of the surreal experience of being behind the scenes of *Big Brother II*: "Even before the housemates were chosen, dozens of technical personnel were being trained in the art of voyeurism" (*Film South Africa*, n.d., 7). A much less spicy variation is *Survivor*, where competitors periodically vote for the excommunication of one among them, in this case based on their (in)ability to survive jungles and wild animals, oceans, and tropical islands. Exposé Television, such as *Cheaters*, records adulterous and kinky behavior; *Temptation Island* sets up and glamorizes promiscuity. Some of these traits of Reality Television have bled into other genres, such as chat shows, spawning crossovers between talk, burlesque, and "true" (*sic*) confessions. The *Jerry Springer Show* is a prime example, featuring errant partners in no-holds-barred public spats, and encouraging brawls involving whole families.

The exposure of dirty little secrets, recorded by cameras and crews that are absent from the screen but present on the sets, seduces audiences into believing that the events being broadcast are real, undirected, and unedited. Unlike narrative cinema, no suspension of (dis)belief is necessary. The basis of this belief, as far as *Big Brother* and similar programming are concerned, is grounded on the promise of banality, scandalous domestic revelation normally swept under the familial carpet, and the illusion of actual rather than enacted deviance, betrayal, and infidelity. Many of these shows promise the *anticipation* of explicit sexual activity, although this seldom actually occurs, and, if it did, the titillation factor would be sacrificed as the show degenerates into pornography, predictability, and impossibility.

Reality Television is often underpinned by the Internet and digital technologies, which create voting channels for *Big Brother*, *Big Brother Africa*, *Pop Idols* and so on. New media have caused astonishing change to popular conceptions of democracy. Reality Television raises questions about what constitutes popular public spheres within specific political and cultural contexts:

- Do these shows impose Western values on global audiences?
- In what ways is Reality Television an indicator of the popular; that is, in what ways are these shows focal points for narratives of everyday issues that concern ordinary people?
- Can *Big Brother* be analyzed as an allegory for Pan-Africanism?
- Is Reality Television a kind of public sphere?

The Democratization of Banality

Where some see Reality Television such as *Big Brother* as a form of democratization of banality involving ordinary people, others demonize the genre as an evil in need of censorship. *Star Academy*, the Algerian version of *Pop Idols*, for example, was declared "un-Islamic" in October 2003 (BBC 2006; Sabbah 2005), and a teenage Moslem participant in South Africa's *Pop Idols* was hounded mercilessly by her religious community in Durban through a "letters to the editor" campaign, amongst other forms of community disapproval. Additionally, *Idols* contestants had to negotiate public sarcasm and insults directed at them by the judges, whose scripts obviously call for commentary that demeans the contestants' self-esteem, self-confidence, and self-identity. An overriding objective of the genre as a whole is found in its competitive value, and its commodification of religious, sexual, and social taboos through the encouragement of individual greed, the rewarding of uninhibited behavior, and enforcement of often personally destructive competitiveness in offering winners the possibility of fame, fortune, and fast-living. *Big*

Brother, too, came in for massive criticism from numerous quarters, and was labeled by Wole Soyinka as "banal, lacking in anything to offer to the Continent" (Jacobs 2006, 8).[1]

It must be kept in mind that three times as many 18–24 year olds voted in the *Pop Idol* final than in the 2001 general election in the UK (*The Guardian*, 9/02/2002, www.guardian.co.uk). Some presidents in Africa would prefer to be the media "idols," but where the pop idols organically mobilize family and community support, some of these presidents enforce support through legislation and brutality, and enact repressive measures to ensure compliant reporting.

What Then, Can the Polity Learn from Reality Television?

Understanding the significance of media begins with an examination of the its social, political, and economic foundations and the cultural construction of its place in society. Media define democracy and help mould the social character of the individual within the discourse of capitalist society. Hardt (2004, 3) notes that "contemporary society is unthinkable without the overwhelming presence of the media, while the never-ending process of mass communication generates a working reality that defines relations among people and events." Others are more forthright, contending that the media are the most important channel of information for people in contemporary society. McNair (2003, 1) states that "[m]odern politics are largely mediated politics, experienced by the great majority of citizens at one remove" through their choice of print or broadcast media. In this view, the mass media become the place for conversation, and ordinary readers, viewers, and voters partake in this conversation only vicariously.

People use media to access knowledge about themselves and the world. At the same time the media deliver audiences to commercial interests, thus generating contradictions in their role and functions. Media practices affect the social settings; including private and public behavior. Media platforms provide education, information, and entertainment. The mass media provide social discourse, and to a significant extent, define, organize, and even determine the communal and political manifestations of social life (Hardt 2004, 3). The media, in sum, are significant in the creation of the contemporary lifeworld.

Nevertheless, the great majority of media enterprises—both within South Africa and globally—are predicated on the model of private, commercial, agendas within unevenly democratic societies that have failed to deliver on their promises of participation. Is it possible, we ask, that the most commercial of the commercial genres and formats—the Reality Television show—has something to offer in the way of civic education, enlightenment, and democracy, which are the very keystones to the democratic functioning of the media, in short, to a mediated public sphere?

The Public Sphere

A vigorously independent media is intrinsic to the infrastructure required for knowledgeable debate, informed choices, and general transparency needed in a mature democracy. This is part of the well-discussed "public sphere"—that space wherein all citizens can freely discuss politics independent of the influence of the state or capital. The concept of the "public sphere" has proved its endurance long after other theoretical ideas have gone out of fashion. Since the work of Habermas was translated into English in 1989, and "discovered" by academics in the English-speaking world, the concept has been applied to media with an almost canonical reverence: it has been developed, modified, critiqued, discarded, and re-embraced with an enthusiasm verging on faith.

For Habermas, the period of early capitalism seemed to approach the "ideal speech situation." During these formative years, discussion among the educated elite was intensely political, focusing on contemporary affairs and state policy. Gentlemen's clubs, salons, and coffeehouses provided the spaces for these informed conversations. The bourgeoisie created networks of information sharing, including newspapers, debating societies, publishing houses, libraries, universities, museums, and the like in order to both express and, more importantly, develop a new political force: *public opinion*.

The concept of the public sphere is most frequently associated with either the liberal print press or public service broadcasting. Discussions of the public sphere within the realm of commercial, entertainment television formats raise a number of conceptual challenges, and require that we go back to the basics and clarify for ourselves precisely what we mean by "public sphere."

The public sphere is a virtual space in which citizens engage in conversations with the guarantee of freedom of assembly and association and express and publish their opinions. These ideals are in line with the ideas of the Enlightenment, which is characterized by four fundamental principles:

- The ideas and their development are independent of Church and State power. They represent the thinking of citizens in a "public" space, and are not dictated by prior dictates or constraints.
- In theory at least, discussion is open to all. In practice, this was not so, eighteenth-century debating circles were the preserve of the wealthy, or at least of the financially comfortable, and they were mainly the preserve of men.

- The public sphere was distinct from private interests—by definition; it served the ideals of the "public," rather than the individual, and of the "citizen" rather than commercial interests.
- The public sphere was driven by a quest for general, universal norms and their rational—that is scientific and objective—legitimation. Thus, discussion and debate within the public sphere axiomatically needs to be governed by the rules of critical reason.

The outcome is a synthesis of ideas, strongly contested but nevertheless raised to a level of acceptance within broader society. Habermas held that these conditions of argument lend public opinion its legitimizing force; public opinion then is distinguished from mere "private" opinion.

Such a schema, by necessity, always was an ideal and somewhat artificial construct. In his analysis of the "structural transformation" of the public sphere, Habermas (1989, 10–11) observed that its erosion was in part due to the growth of the advertising and marketing industries. In other words, while it was the development of widespread capitalism that engendered the public sphere, it was precisely the growth of capitalism that proved to be its greatest nemesis. Habermas contended that media commercialization excluded political questions from large areas of the public sphere. In particular, he pointed to the tendency toward the consolidation of media ownership, and the move toward monopolies, or quasi-monopolies, resulting in the uneven distribution of wealth, the rising cost of entry into the media and other forums facilitating the public sphere, and the resultant unequal access to/control over the public sphere (Habermas 1989, 10–11). Whereas in the idealized concept of late-Victorian "public life," "publicity" connoted openness in which the affairs of state were exposed to public scrutiny, the modern conception of "publicity" has come to mean those aspects that attract attention for the purposes of journalism, advertising, and politics. In such situations, the commercial control of media stems from the flow of "public" information, just as the continued dominance of the state over many forms of African media continue to do.

A three-tiered view of the public sphere, the *micro*, *meso* and *macro*, is proposed by Keane (1996). While largely descriptive rather than conceptual, Keane's schema offers a useful way of applying the more abstract notion of the public sphere to concrete situations. *Micro* public spheres are the small-scale interactions between dozens, hundreds, or even thousands of people at subnational levels. An example is found in the Botswana *kgotla* (Kerr 2001), and the South African *imbizo* (cf. Mabelebele 2006) and *bosberaad*. Significantly, it is often at the *micro* level that social movements use different media to question and transform the status quo. The next level is the *meso* sphere, where millions interact at the national level.

The *meso* level also refers to communication between neighboring states or between regions within a state. At the *macro* level, communication takes place between hundreds of millions at the supranational or global level. The development of the *macro* level is located in processes that emerged after the demise of the cold war from 1990 onward. DStv, for example, connects South African and African viewers to global news channels. The schema is useful in discerning the levels on which *Big Brother* has acted as an African public sphere.

Some Structural Issues in Africa's Changing Mediascape

It is worthwhile reminding ourselves of the authoritarian tendencies apparent in African media across the continent. Almost all broadcasting, and the greater part of the print media, historically were part of the state structures and, as such, were under the authority of the government in power (see Lush 1995). This government, typically though not always, was a one-party state headed by a powerful leader. Consequently, the connection between the state leadership and the media were often extremely close, as attested by the previous president of Namibia, Sam Nujoma, who held the offices of president and minister of broadcasting simultaneously (Heuva 2007, 127).

We will not delve into the changes in the print media in Africa (cf. Nyamnjoh 2005); suffice it to remind ourselves that despite their small circulations and even smaller penetration into rural areas, the print media remain politically influential, as they are the touchstone of the ruling elites, both in government and in business. Growing liberalization throughout the arenas of media, and civil society in general, has meant that the traditional state-owned newspapers are now in competition with commercial ventures, some of which are funded by global corporate and donor interests (see Deane et al. 2004, 179–180; Kasoma 1997). Yet, other than in South Africa, inadequate advertising base for independent commercial media remains an impediment to any serious growth in media autonomy.

The opening up of the media as a result of liberalization has not been without a major clawback by governments (Heuva et al. 2004). The 1991 Windhoek Declaration was the first response to new, enlightened regulatory processes. The mere availability of technology becomes an issue in Africa (Mjwacu 2002), especially when it is subjected to censorship. For example, in 2001, during the month marking the tenth anniversary of the Declaration's publication, there were several ironic developments including the resignation of the first head of the new Botswana Television over a government instruction not to screen a documentary on capital punishment on its just introduced television service (Mosime 2007, 168), and the arrest of Malawian vendors selling newspapers that published columnists critical of the president's desire for a third term. These somewhat arbitrary developments point to the realization that, while real strides have been made in providing strong, independent voices to counterpoise government propaganda, commercialization in itself has been insufficient to ensure true political independence.

Broadcasting

In broadcasting, radio remains the preeminent medium throughout Africa. It has the widest spread; it is relatively inexpensive to produce and to receive; it does not require a literate audience; and because much radio is broadcast in local languages and dialects, it is available to all strata of society, both urban and rural, and across all economic classes. Even more than print, radio has benefited from liberalization, with a burgeoning number of community and rural stations coming on air, along with a new generation of commercial FM stations. The downside of this scenario is that many of these recently licensed commercial stations, dependent entirely on advertising revenue for their sustainability, eschew news and current affairs for more popular music formats. When they do air news, it is often heavily dependent on, or even supplied by, international agencies such as the BBC, Reuters, or Agents France Presse (see Deane et al. 2004; Mytton et al. 2005). As platforms from which to promote a public sphere, they are of limited use.

Television in Africa, with a few exceptions, was introduced after each of the new states had achieved independence, and at a time when government monopolies over the electronic media were well established. TV is less developed in Africa than in any other continent. Fewer people have a TV set at home and there are fewer TV stations transmitting to the population than on any other continent.[2] TV in sub-Saharan countries, like radio, is dominated by state-owned and state-controlled broadcasting. However, unlike radio, TV is almost entirely national, rather than local or regional, in character. It is dominated by imported content, mainly from Europe and the United States. Whereas radio makes good use of local languages, TV is dominated by three European languages: French, English, and Portuguese; and also Arabic in North Africa. Finally, and crucially, chronic underinvestment and poor facilities have lead to low levels of local production. Despite the dearth of anything approaching "local content," however, the levels of inter-African cooperation in TV production, exchange, and marketing remain low (Mytton, Tusque, and Teer-Tomaselli 2005).

Traditionally, broadcasting in Africa has been part of the civil service. It is this top-down manifestation of communication, in which governments speak "to," rather than dialogue with, their citizens, which characterizes communication systems across the world. Some states and governments continue to restrict broadcasting autonomy, and in many instances, radio and TV continue to be used as the propaganda arms of the ruling elites. Since 1990, however, the democratizing and privatizing impulses stemming from a number of different sources have spurred the region's governments into reregulating the airwaves. Private satellite transmission is now available in many countries in addition to terrestrial services and commercial channels.

Malawi and Botswana introduced fledgling TV services in 2000, while in all the African countries, issues of freedom of the press and conflicts between state and the private press have been reignited by deregulation and reregulation policies, in concert with liberalization of national economies (Mbay'o et al. 2000; Kasoma 2000, 1997; Tomaselli and Dunn 2001). In South Africa, the free-to-air e-TV, owned in part by Hoskins Consolidated Investment (HCI), and financed in turn by Trade Union Pension Funds, is openly critical of Zimbabwe in its newscasts, while the South African Broadcasting Corporation (SABC) felt the political heat when it carried negative reports that undermined President Mbeki's inexplicable "quiet diplomacy" as that country disintegrated.

As has been the case with print, liberalization does not necessarily mean the end of state control, or quasi-state control. The way reregulation and privatization occurs affects the outcomes. Nation states ensure compliance of their regulations through a variety of means. Zimbabwe nationalized broadcasting to the state. Namibia made SWAPO the main shareholder of Namibian Broadcasting (Heuva 2007, 10–11). Botswana, South Africa, and Zimbabwe have, at various times, influenced personnel appointments and forced resignations. In extreme cases, where the press is concerned, bombing, outright censorship, excommunication, torture, and detention of journalists and editors have become the norm.

Some Characteristics of African Media under Late Capitalism

In this specifically digital age, there are a number of mutually reinforcing and interlocking processes. *Globalization* is the most crucial of these. Globalization within African media occurred when foreign interests purchased shares in the domestic media; and domestic firms, especially those of South Africa, Zimbabwe, and Nigeria, purchased international interests. The producer of *Big Brother Africa*, for example, was Endemol, a Dutch-based multinational, with a 50 percentage South African black "empowerment" shareholding, and offices at DStv, Johannesburg. Where governments fear an open public sphere, decrees made to curtail criticisms of government also restrict foreign ownership in domestic media, and indeed, domestic ownership in foreign companies, such as the Zimbabwean-owned *Mail and Guardian* based in South Africa. In contrast, in 2003 South Africa relaxed its foreign ownership restrictions on telecommunications and broadcasting as it tried to raise adequate international investment capital to improve both urban and rural services. Global infrastructural ownership and investment often keeps open certain sectors of the public sphere that national governments want to close down. The *Mail and Guardian* is critical of the government in its country of corporate citizenship, for example. This is why authoritarian governments try to prevent their citizens from interacting at the *macro* level. In a healthy democratic state, communication should occur freely across all three levels.

A school of thought, dubbed the "hyperglobalizers" (Boyd-Barrett and Rantanen 1998, 5), considers globalization as a process that ends the concept of the traditional nation states. This school suggests that national governments need to get used to such an infrastructure, like the post-modern public sphere, which disregards the borders of the traditional nation state. "The old dominance of state-structured and territorially bounded public life . . . is coming to an end. Its hegemony is rapidly being eroded by the development of a multiplicity of networked spaces of communication . . . not tied immediately to territory" (Keane 1996, 34). While such expressions of faith seem to be a quaint overstatement a decade later, and the effects of globalization on the continuation of national sovereignty have been exaggerated, nation states do find themselves somewhat limited in the extent to which they are able to control cultural artifacts from beyond their borders. In this regard, the expansion of transnational TV throughout Africa, limited as it is to the wealthy elites in the urban areas, is an instructive development.

Transnational TV and the Establishment of DStv in Africa

Transnational TV in Africa is extremely limited. In contrast to radio, which crossed national boundaries from its earliest days, TV, for many years, was broadcast within the boundaries of particular countries. However, fuelled by advances in TV technology, together with a relative liberalization of the media and the phenomenon of globalization, by the turn of the new millennium a small degree of transnational and international TV penetration had occurred. For transnationalization to happen, changes in the political and cultural sphere were necessary. The more liberal atmosphere necessary for transnational TV to grow was slow to develop in many parts of Africa. To some extent it is still to emerge in some countries. The same impulses of liberalization that allowed for the emergence of several hundred private radio stations and the retransmission of international shortwave services on local FM frequencies have opened the market to the wider distribution of satellite TV services, both private and state funded. With an annual monthly subscription of approximately US$60, the reception of these services is limited to foreign nationals and the local elite, and thus they do not contribute to the enrichment of the local states, either culturally or economically.

It is against this background that we should view the emergence of MultiChoice Africa, as well as a number of other satellite services, mainly from Europe and the Middle East. MultiChoice is the preeminent content carrier in Anglophone Africa. The parent company, MultiChoice Holdings Investments Limited (MIH), in Johannesburg, operates subscription services across Africa, as well as the Middle East, Greece, and Thailand. It emerged from M-Net, South Africa's first private TV channel (cf. Collins 1992).

An analogue service, distributed via satellite, was launched to more than twenty African countries in 1992. The first transnational joint venture was entered into with Namibia in 1993 (Heuva 2007). Digital satellite services were offered across Africa in 1995, utilizing the C-band on PAS4 satellite. With delivery via satellite, the bouquet of channels was enlarged significantly. This was the beginning of DStv (Direct Satellite TV), a subsidiary of MultiChoice. At this point, M-Net and MultiChoice were formally separated, with the latter changing its name to MIH. During the next few years, MultiChoice expanded beyond Africa into Europe and the Far East. In 2000, the company launched Indian and Portuguese bouquets on the DStv satellite service in southern Africa. Thus within a single decade, MultiChoice expanded horizontally, moving from being a terrestrial pay-TV platform in a single country to a multi-platform provider across the African continent. In 2006 the group claimed over two million paid up subscribers in Africa, the Mediterranean, and Asia, 1.25 million of whom are in Africa. Of the African customers, 57 percent subscribe to the digital platform (DStv): 705,000 households in South Africa and a further 251,518 in the remainder of sub-Saharan Africa. The analogue service accounts for 339,422 households in South Africa, and 12,798 in the rest of Africa (source: www.multichoice.co.za/ourbusiness).

Content and Local Production

There is something of a gulf in sub-Saharan Africa between the number of TV stations available on satellite and the number of channels actually seen by African audiences. In the former case, it was calculated that, in July 2002, around 250 channels were transmitted by some twenty-two satellites over Africa, of which approximately 100 were in English, 86 in French, 30 in Arabic, seven in Portuguese, several in major Indian languages, and the others in different non-African tongues (see Mytton,

Teer-Tomaselli, and Tudesq 2004). Only three were in sub-Saharan African languages—one in Amharic and two in Afrikaans.

A major characteristic of transnational TV in Africa is that, until recently, rather little of its content is specifically designed for or created in the continent. Many of the present transnational TV services received in Africa are actually produced for domestic channels in Europe or North America—for example for TF1 or Canal + in France, the BBC in the UK, or NBC in the USA. Others are produced for international audiences in all parts of the world, such as most of the output of CFI, CNN, and BBC World. Aside from some South African production that can be seen on some services available from that country, there is at present not very much else on offer that can be described as indigenous. *Big Brother Africa* was syndicated on national and commercial broadcasters across Africa, as well as being shown on DStv.

MultiChoice Africa (MCA) is a broadcaster-publisher and subscription manager. In its former guise, MCA packages channels, some fully imported from America, Europe, or Asia, others compiled from imported programming together with locally commissioned programming through its sister company, M-Net, and broadcasts these as channel "bouquets" to subscribing clients. It was M-Net that produced the first transnational content for satellite, utilizing the Endemol format, *Big Brother 1* (2001). The formula proved so successful that in 2003 the project was extended to the whole of Africa under the title, *Big Brother Africa*, with an all-African cast vying for the final prize. *Big Brother Africa* is a manifestation of this effort toward transnationalization. While South African material has been well received throughout Africa, there was a need felt to produce a program that would act as "local content" for the whole of Anglophone Africa, hence the "national" character of the reality format was substituted for a transnational version. To maximize the commercial potential of the show, a partially interactive channel was dedicated to it. In this way, the public sphere at which it was aimed was changed from the *micro* to the *meso* scale.

The house for *Big Brother Africa* (2003), located in Johannesburg, substituted a restricted media-based democracy in an artificial display of African unity and adherence to democratic ideals. *Big Brother Africa* was then the largest African reality show, with competing contestants from twelve countries (Ghana, Kenya, Uganda, Tanzania, Malawi, Zambia, Angola, Namibia, Botswana, Zimbabwe, Nigeria, and South Africa). Daily viewership of this twenty-four-hour transmission on two DStv channels was estimated to be between twenty-five and thirty million (www.bigbrothersa.com). Viewers put aside rivalries by occasionally voting against their own national representative in favor of a contestant from another country. What *Big Brother Africa* revealed was the way in which

Reality TV works across the three levels of public sphere:

- Individuals, supported by communities at the *micro* sphere, engaged in a competition that stimulates interaction at the *meso* (national and regional) level.
- Then, when one contestant was "swapped" with a housemate from the *Big Brother* show in the UK, *Big Brother Africa* was aired at the *macro* (global) level.

The show's formula was sold to countries worldwide, making the housemates interactions (micro) both national (meso) and global (macro). Similarly, *Pop Idol* finalists are recruited from auditions across the country (micro), they compete against other nationals (meso), and the winner then competes at the international level, where viewers from all the countries can influence the vote (macro).

Big Brother constructs artificial "tribes" at the micro level, broadcasting to a macro audience that simulates the fast-diminishing eighteenth-century coffeehouse public sphere as originally conceptualized by Habermas (1989). While abstract, symbolic, and actual power remains under the control of governments, Reality TV, controlled by producers, facilitates the illusion of participatory democracy. Electronic interactivity confers upon viewers the power to vote individual contestants off the program. In another Reality format, the talent show *Pop Idol*, an example of "the worst excesses of capitalism and democracy gone mad" (Des Lindberg, www.3rdearmusic.com), gives the audience "electorate" the illusion that they are in control of the judging process. This is achieved either through the use of mobile telephony, which has registered phenomenal expansion throughout Africa (cf. Teer-Tomaselli 2006), as well as through Internet-enabled e-mail. However, the voting system is skewed fundamentally in that it is geared toward the affluent who subscribe to DStv, have access to Internet services, and who are able to afford calls to and from cell phones. It also works in favor of contestants who come from large families and close-knit communities, and are thus able to mobilize the greatest (cellular) support, irrespective of their talent. During *Big Brother 1* (2001), newspapers in South Africa made these shows and their bizarre characters their lead stories, and competitions and phone-ins mobilized consumers in ways that baffle even the most interactive of mature democracies.

The separation of the public and the private was collapsed as individual greed and vicious interpersonal competition replaced real news and real politics, which, for many viewers, were boring or irrelevant. The import of these circumstances was not lost on "real" (as opposed to "Reality") politicians. Blair starred in *The Simpsons*, the satirical animated TV series, in what may have been an effort to shake his stuffy image (Brant 2003), and former New York mayor Rudi Giuliani acted in cameo parts in two U.S. programs, a sitcom (*The Cosby Show*) and a police drama (*Law and Order*). Bush metaphorically portrays every American screen hero in his TV appearances. Actors become presidents, as in the Philippines, politicians become actors, celebrities get votes, and new nouns are born—the Terminator becomes "'The Governator': starring in a legislature near you."

The question is, what can advocates of democracy learn from the *Big Brother* experience? How can the polity be made as sexy and necessary as Reality TV has become? On February 16, 2004, "Big Bad Brad," a brash contestant in *Big Brother I* even launched his own ill-fated political party to contest the 2004 South African elections. *The Big Brother Africa* winner from Zambia was awarded a diplomatic passport to act as an ambassador for her country.

Freedom of Information and the Public Sphere

Most of the structural processes impacting national media throughout Africa have occurred within a policy vacuum, with the single exception of South Africa. The result has been a lack of transparency and public accountability in most states. Only South Africa has a civil society and union movement strong enough to sustain the social dialectic.

In South Africa, the Promotion of Access to Information Act (PAIA) (2000) supposedly promotes transparency and accountability. However, the Act as published is largely a watered down version of the original draft, especially as the new government realized that while access is good for democracy, democracy needs to circumscribe access—if anyone is able to pry unrestrictedly on any topic into the behavior of public officials and state institutions especially, cover-ups become more difficult. Also, national security considerations may apply. *Big Brother*, like freedom of information legislation, creates the illusion of reality and free access, but both forms of information facilitation are shaped, conditioned, and orchestrated in a variety of ways invisible to readers/viewers/citizens.

Big Brother has become a hook on which some critical journalists have discussed issues of the polity. Questions we could pose in this regard could include:

- Do hypermediated postrevolution African generations relate more to Reality TV than they do to their own material conditions?
- Or do consumers perhaps make better sense of themselves as citizens in and of the world through the genres of Reality TV?
- Is it possible that *Big Brother Africa* did more to popularize pan-Africanism than did all of Nkrumah's monuments and publications, Mbeki's African Renaissance speeches, and Gaddafi's bank account?

For instance, "good taste" reporting required by Kasoma (1997), linked to the discourse of (im)morality, often serves to conceal the much more pervading structure of patriarchy that entraps entire societies. Perhaps *Big Brother Africa* offered a form of indigenization: it struck a chord with tens of millions of African viewers; generated a transnational debate on who is an African; and broke nationalist voting patterns. It might have been "bad" entertainment in terms of conventional middle-class morality, but it was "good" entertainment by intersecting with popular concerns.

Politics of Power

The "public sphere" is an arena between the power of the state or government, on one hand, and the power of the commercial and industrial base, on the other. People come together in this arena as citizens, rather than as political players or motivated solely by commercial interests. Emphasis is placed on receivers as citizens, not only as voters or consumers. The central importance of the mass media in the functioning of political and social life has come about as a result of the decline of other *forums* for interaction, particularly that of the political meeting. The privatization of the media, and its highly visual nature, has meant that politics is placed alongside popular culture, the result being that celebrities are often foregrounded as the new leaders. In the UK, where this trend is more pervasive, politicians are aware of this and are concerned about voter apathy. Endemol entered into talks with British broadcasters about creating a Political Pop Idol, making citizens consumers of the political process in order to create the nation's first genuinely populist politician (Hinsliff 2004). In Columbia, President Uribe appeared on the local version of *Big Brother* to speak about his proposed austerity measures in the run-up to the national referendum. The show's producers said his appearance was "a public service" deemed necessary after polls showed that Columbians did not understand the policies under review. The *Big Brother* housemates were given the task of discussing the referendum instead of the usual tasks given to them, such as striptease competitions and nonstop dancing.

Kasoma (2000) divides journalism into "bad" and "good." He argues that scornful and disparaging journalism opens up the print media especially to state interference. But what one person or reading community considers "bad" may be considered "good" by another. Obviously, there are limits: Chinweizu (1999, 368–370), from Nigeria, identifies the Seven Deadly Sins of Media Vandalism, among them are "liespapers," "junk," "quisling," "payola," "guerrilla journalism," "freedommania," and "unfairness." That these kinds of "scornful journalism" practices may negate dialogue is accepted, but satirical, ironical, and tabloid journalism does have a role in testing the boundaries of critique. These genres, which have their own internal rules of logic and narrative, are therefore fundamental to maintaining the discursive contours in any democracy. Perhaps, "bad" journalism is also a kind of genre—especially in our late modern world where conventional rules about "truth," a recurring subject in Kasoma's studies (cf. Kasoma 1997, 299), no longer apply quite so absolutely. If journalism is a "factual form of realism,"[3] and readers/ audiences are understood to be codeterminers of meaning, then what is "good" and what is "bad" becomes quite difficult to assess. It is in this intersection that Reality TV comes into its own. It creates its own truth, structured in terms of production strategies, genre manipulation, and audience preferences.

The *pastiche* that characterizes "postmodern" forms of cultural production, including Reality TV shows, is evident in the diplomatic status conferred by Zambia on its *Big Brother Africa* contestant. Reality and truth are no longer pre-given. The content of truth now is largely negotiated, and "reality" arises out of social negotiations. Hard-and-fast facts are always subject to interpretation from the vantage point of different paradigms, ideologies, and different systems of logic. The only thing that is "real" is which interpretation or paradigm is commonsensically agreed upon at any moment. However, both the hegemonic meaning and moment can change in an instant. Absolutes seem difficult to identify.

"Good" and "bad" journalism are two sides of the same coin. Dialogue in a functioning public sphere occurs in between and within these discursive and contradictory fissures. So, perhaps within Africa, where "a large part of the population does not and have no possibilities of taking part in such a process of scrutiny, nor be informed about it" (Rønning 1997, 21), *Big Brother Africa* and other programs like it, for all their gutter sensation, may contribute in some way to the democratic process after all. They mobilize audience (ultimately comprised of *citizen's*) interest; encourage mass (virtual) participation (both voyeuristic and voting); and reward socially preferred patterns of choice (via electronic voting). However, these practices do not facilitate democratic conversations (cyber-democracy) and political participation (cyber-government) among viewers. They do not encourage debates about general public opinion, the public sphere, or the media per se, except by journalists (and academics), and sometimes by the producers also who tend to use cultural studies jargon of "the popular" to defend their programs.

Although often used interchangeably, cyber-democracy and cyber-government (also referred to as e-government) refer to two separate procedures, as indicated above. E-government allows for the electronic administration of government tasks; it is "the use of digital technology in the management and delivery of public services, predominantly through the internet" (Edmiston 2003, 20). Unlike this largely administrative function, e-democracy highlights "the interactive nature of Web sites, arguing that government sites should be more than electronic bulletin boards which simply provide information" and maintaining that "website design should encourage interaction with government officials and provide opportunities to provide input into the decision-making process" (Ferber, Foltz, and Pugliese 2004). However, cyber-democracy can also lead to the fragmentation of political views into polarized sectors, which "filter out unwanted information, tailor their own news and congregate at specialized Web sites that closely reflect their own views" (Stille 2001) through the use of discussion boards and the creation of exclusive web-groups or Web sites.

This heterogeneity of audiences, created, and exacerbated, by new media technologies and globalization, fractures any sense that they can be uniformly "represented" by the media or state, and assumptions that journalism, as an institution, must automatically respect "authority" no longer apply. Cyber-democracy and the Internet should be seen as a site of struggle for power and as a contested terrain. Dominant corporate and state powers, as well as conservative and rightist groups, use new technologies to advance their agendas. They can also be used to advance a progressive agenda and the interests of the oppressed, and the forces of resistance and struggle.

Some journalists and late-night show hosts used the genre of satire to lampoon reality shows. Satire is sometimes also used by politicians to discredit their critics. Satire is a kind of "cultural ownership" that opens up a channel of communication in communities with opposing values. In so doing, satire can liberate the minds of people who are restricted by social norms (Bakhtin 1984). These "unofficial" discourses allow for the exchange of opinions and prise open communication between social norms and minority views. Comedian Dave Lippan describes humor as having an "unmasking" function that "brings society's rulers down to the level of the populace, thereby facilitating democracy" (www.towardfreedom.com). Political jokes and satire are often used for airing grievances that are not given an "official" outlet (Powell and Paton 1988). While such subversive discourses may cause offence at the governmental level, they are essential to the effective functioning of a democracy, which is reliant on an informed populace who are able to engage in the free exchange of ideas.

Conclusion

Debate, dissent, argument, disagreement, and criticism are the motors of constructive and democratic progress and development. The Frankfurt School taught that to kill the social dialectic in any society is to terminate democracy and justice, fought for over generations, even centuries. The Internet has impacted the four components that make up the public sphere, namely the news media, conversations, public opinion formation, and participation, and in many aspects they have indeed improved parts of the public sphere.

Even though we may agree that Reality TV as a genre is not particularly educative or culturally enriching in terms of the norms of "high culture" or so-called African values, the genre is popular worldwide and is serving some kind of public sphere purpose for its audiences. One South African columnist wrote of *Big Brother* that *Big Brother* is not watching us—we are watching *Big Brother*. So extraordinary was the viewership during its first week of transmission in September 2001 that the national Internet infrastructure all but collapsed because of its overuse by viewers wanting to access the show's Web site. In one village in Namibia, neighbors banded together to buy DStv so they could watch *Big Brother Africa* communally. When Namibian President Nujoma jokingly appeared to want to take the show off Namibian TV, one viewer called a local radio station to say she would no longer be voting for the ruling party (www.bigbrothersa.com). The power of the market has become the market of power. Governments perhaps become somewhat irrelevant in this new set of relations. Some of their efforts are simply last gasps at retaining national boundaries in an age of porous borders.

Notes

Thanks to the Francis Peter Kasoma Foundation, which invited a much earlier version of this paper at its inaugural lecture in Lusaka, 2004. Thanks also to Anusha Govender, Wendy van de Weg, Deidre Donnelly, and Elva Gómez De Sibandze for their research and editorial assistance.

1. The version of Sean Jacobs' paper from which all the quotations are taken is the original English language draft. Page references in the published editions differ. Published version: Jacobs, S. 2007. "Big Brother, Africa is Watching." *Media, Culture and Society* 29(6): 851–868.
2. Accurate and reliable statistics for Africa are hard to find. The BBC's international research department estimated in 1999 that there were 29 million television sets in sub-Saharan Africa, about one for every twenty people compared to Asia and North Africa where there was one set for every six people. See G. Mytton, "From Saucepan to Dish," in, *African Broadcast Cultures*, ed. R. Fardon and G. Furniss (Oxford: James Currey, 2000). As far as TV transmitting facilities are concerned, there are several African countries with only one terrestrial domestic broadcaster. In many countries, satisfactory reception can only be obtained in and around the major urban areas. See *World Radio TV Handbook* (Oxford: WRTH Publications, 2004).
3. Statement by John Hartley at the 2004 Crossroads Conference, during a plenary session.

References

Bakhtin, M. 1984. *Rabelais and his world.* Indiana: Indiana University Press.

Boyd-Barrett, O. and T. Rantanen. 1998. *The Globalization of News.* London: Sage.

Chinweizu. 1999. "Towards the African Renaissance Media." In *African Renaissance*, ed. M. W. Makgoba. Cape Town: Tafelberg.

Collins, R. 1992. "Broadcasting Policy for a Post-Apartheid South Africa." *Critical Arts* 6(1): 26–51.

Daniel, J. (n.d., no additional information) *Film South Africa.*

Deane, J., K. Dixit, N. Mue, F. Banda, and S. Waisbord. 2004. "The Other Information Revolution: Media and Empowerment in Developing Countries." In *Communicating in the Information Society*, ed. B. Girard and S. Siochrú. Geneva: United Nations Research Institute for Social Development.

Edmiston, K. D. (2003). "State and Local E-government Prospects and Challenges." *American Review of Public Administration* 33(1): 20–44.

Ferber, P., F. Foltz, and R. Pugliese. 2004. Cyberdemocracy v. E-government: The Degree of Interactivity on State Legislative Web sites. Paper prepared for delivery at the 2004 Annual Meeting of the American Political Science Association, Chicago, Il., September 1–5. http://java.cs.vt.edu/public/projects/digitalgov/papers/Ferber.Cyberdemocracy.pdf.

Habermas, J. 1989. *The Structural Transformation of the Public Sphere: An Inquiry into a Category of Bourgeois Society.* Cambridge: Polity Press.

Hardt, H. 2004. *Myths for the Masses.* Malden, MA and London: Blackwell.

Heuva, W. 2007. The Political Economy of Broadcasting and Telecommunications Reform in Namibia: 1990–2005. Unpublished PhD Thesis, University of KwaZulu-Natal.

Heuva, W., K. G. Tomaselli, and R. E. Teer-Tomaselli, 2004. "The Political Economy of the Media in Southern Africa." In *Who Owns the Media. Global Trends and Local Resistances*, ed. P. N. Thomas and Z. Nain. London: Zed.

Hill, A. 2002. "Big Brother: The Real Audience." *Television and New Media* 3(3): 323–340.

Jacobs, S. n.d. Big Brother, Africa is Watching. Mimeo.

Kasoma, F. 1997. "The Independent Press and Politics in Africa." *Gazette* 59(4–5): 25–310.

———. 2000. *The Press and Multiparty Politics in Africa.* Tampere: University of Tampere.

Keane, J. 1996. "The Structural Transformations of the Public Sphere." In *Media and Democracy*, ed. M. B. Anderson. Oslo: University of Oslo.

Kerr, D. 2001. "Media Democracy in Botswana: The Kgotla as Myth, Practice and Post-Colonial Communication Paradigm." In *Media, Democracy and Renewal in Southern Africa*, ed. K. G. Tomaselli and H. Dunn. Colorado Springs: International Academic Publishers.

Lush, D. 1995. "Namibia." In *Who Rules the Airwaves? Broadcasting in Africa*. ARTICLE 19 and Index on Censorship: London.

Mabelebele, J. 2006. "Ideological Objectives Underpinning *Imbizo* as a Model of Communication and Governance." *Communicare* 25(2): 103–125.

Majwacu, T. 2002. "Opportunities and Challenges of New Technologies in Media and Communication: The Windhoek Declaration." *African and Asian Studies* (14): 407–428.

Mathijs, E. and J. Jones, eds. 2004. *Big Bother International: Formats, Critics and Publics*. London: Wallflower Press.

Mbay'o, R. T., C. Onwumechili, and R. N. Nwanko, eds. 2000. *Press and Politics in Africa*. Lewiston: The Edwin Mellon Press.

McNair, B. 2003. An Introduction to Political Communication. London: Routledge.

Mosime, S. 2007. Botswana Television BTV Negotiating Control and Cultural Production in a Globalizing Context: A Political Economy of Media State Ownership in Africa. Unpublished PhD Thesis, University of KwaZulu-Natal.

Mytton, G., R. E. Teer-Tomaselli, and J-P. Tudesq. 2005. "Transnational Television Worldwide: Towards a New Media Order." In *Transnational Television Worldwide Globalisation and Transnationality in the Media*, ed. J. Chalaby. London: IB Tauris.

Nyamnjoh, F. 2005. *Africa's Media: Democracy & the Politics of Belonging*. South Africa: UNISA Press.

Powell, C. and G. Paton. 1988. *Humor in Society*. Basingstoke, Hants: Macmillan.

Rønning, H. 1997. "Institutions and Representations." In *Media, Democratization and Identity*, ed. R. Zhuwarara, K. Gecau, and M. Drag. Harare: University of Zimbabwe.

———. 1998. "Democratization Processes in Southern Africa and the Role of the Media." In *Perspectives on Media, Culture and Democracy in Zimbabwe*, ed. R. Waldahl. Oslo: IMK Report No 33.

Tomaselli, K. and H. Dunn, eds. 2001. *Media, Democracy and Renewal in Southern Africa*. Colarado Springs: International Academic Publishers.

Tincknell, E. and P. Raghuram. 2002. "Big Bother: Reconfiguring the 'Active' Audience of Cultural Studies?" *European Journal of Cultural Studies* 5(2): 199–215.

Teer-Tomaselli, R. 2007. Lessons from the Transformation of South African Broadcast Media Post. Unpublished Inaugural Lecture, Durban: University of KwaZulu-Natal.

Internet Sources

Accueil (n.d, no other info). Internet and the Public Sphere: Cyber Democracy and Cyber Government. http//www.vecam.org.

BBC. 2006. Pop Idol Pulled from Algerian TV. http://bews.bbc.co.uk/2/hi/africa/4696866.stm.

Berger, M. n.d. Kiss My Aztec: The Functions of Satire. www.rohan.sdsu.edu/dept/drwswebb/3_1/berger.pdf.

Brant, R. 2003. Blair's Simpsons Role is Great PR. *BBC News*. http://news.bbc.co.uk/go/pr/fr/-/1/hi/entertainment/reviews/3234778.stm.

Lindberg, D. 2002. Idle Democracy Gone Mad. www.3rdearmusic.com/forum/mediapow.html.

Lippan, D. n.d. "Cheap Shots: How the Kings of Late-night Avoid the Issues." In *Toward Freedom*, online magazine. www.towardfreedom.com.

No author. Curtain Falls, the Show Lives On. www.bigbrothersa.com.

No author. 2002. Ebden, Abrahams Drive Flashy New Cars. http://iafrica.com/news/sa/19717.htm.

No author. 2003. Arab Idol a Battle of Nations. *CBS News*. www.cbsnews.com/stories/2003/08/18/world/main568959.shtml.

No author. 2003. Colombia President on Big Brother. *BBC News*. October 1. http://news.bbc.co.uk/go/pr/fr/hi/entertainment/tv_and_ra dio/3154876.stm.

No author. 2004. Arab Big Brother Show Suspended. *BBC News*. March 1. http://news.bbc.co.uk/go/pr/fr/-/2/hi/middle_east/3522897.stm.

Sabbah, H. 2005. Saudi Religious Scholars Condemn Arab Pop Idol. http://sabbah.biz/mt/archives/2005/05/12/saudi-religious-scholars-condemn-arab-pop-idol/.

Newspaper Sources

de Silva, N. 2004. "His Master's Voice?" *CPU News*. February, 4.

Hinsliff, G. 2004. Power Idol? Now Politicians Audition for Votes. *The Observer*. January 4. http://observer.guardian.co.uk/politics/story.

Khan, F. and Hartley, A. 2003. Debate on "Forbidden" Idols Gathers Steam. *Daily News*. October 9, 1.

No author. 2002. Will or Gareth? *The Guardian*. February 9, 2002. www.guardian.co.uk.

No author. 2004. Arab Big Brother Upsets Islamists. *The Sunday Times* (Australia). www.news.com.au.

No author. 2004. Arabs Kick Out Big Brother. *The Australian*. March 3.

Staff reporter. 2003. Idols Show Forbidden, Council Tells Muslims. *The Cape Argus*, October 7, 1.

Staff reporter. 2004. Bahrain MPs Seek to Grill Minister on Reality TV Show. *Gulf News*. February 26. www.gulf-news.com.

Stille, A. 2001. Negative Effects of Cyberdemocracy: Adding Up the Costs of Cyberdemocracy. *New York Times*, June 2. http://query.nytimes.com/gst/fullpage.html?res=9406E7DB163FF931A35755C0A9679C8B63.

CHAPTER 12

Popular Music, New Media, and the Digital Public Sphere in Kenya, Côte d'Ivoire, and Nigeria

George Ogola, Anne Schumann, and Michael Olutayo Olatunji

In many parts of Africa, successive governments have historically either monopolized or manipulated public spaces for popular expression. New "spaces of freedom" have consequently been forged and fashioned around popular cultural forms. Denied space in public sites of communication, alternative narratives have found expression in popular music, dance, drama, and other formats. These popular cultural forms often provide spaces for political engagement and remain strategic spaces for dissent. But these forms are not exclusively a province of either the subaltern or of subversion. In many countries, they are exploited by both the subaltern and the potentate. Thus shared, they simultaneously contest domination and affirm it. Indeed, it is this dialectical but contradictory relationship that informs concepts such as Achille Mbembe's idea of "mutual zombification" in the postcolony. From a comparative perspective through our study of *Yabis* music in Nigeria, *Zouglou* in Cote d'Ivoire, and variants of hip-hop music in Kenya, such as *Genge* and *Kapuka*, we seek to demonstrate how popular music articulates political processes and how musicians now exploit the opportunities new media provides. The exercise of power in Africa is in many ways a performance, one contested and affirmed best in a popular social grammar lived by the subaltern and captured most eloquently by popular music.

Introduction

The (re-)introduction of political pluralism in several African countries in the 1990s allowed for the development of a relatively non state-dominated media infrastructure. This period saw the establishment of new FM stations and private TV stations as well as increased access to satellite channels, subsequently providing space for the growth of local cultural forms such as urban popular music. Largely located within Africa's conurbations, urban popular music gradually became the legitimate site for reflecting and engaging with the anxieties and dilemmas of Africa's postcolonial existence. This chapter explores popular music as a platform for political and social engagement in Kenya, Cote d'Ivoire, and Nigeria. However, we go beyond the generic and problematic narrative that often focuses on the dissentious nature of popular music. While acknowledging its oppositional disposition, we also problematize our reading of the "popular," emphasizing its complex, even contradictory, character, indeed its ability to counter hegemonic discourses but also to be coopted by statecraft. Crucially, we also note the genre's dynamism especially within the context of Africa, a continent described by Mbembe and Nuttall (2004, 351) as "a space of flows, flux, of translocation, with multiple nexuses of entry and exit points, (a place) fundamentally in contact with an elsewhere." To this end, we agree with Ranger (2005) that Africa and its cultural forms need to be interrogated in terms of questions of circulation, flows, and transnationalism (cited in Ogude and Nyairo 2005). Consequently, the key question is: How has new technology impacted on the growth and circulation of popular music, especially in engagement with either the mediation or shaping of sociocultural and political processes?

The first section of the chapter interrogates how Kenyan urban music articulates the country's social and political order. It also demonstrates how new media enables the appropriation of transnational discourses on gender, identity, politics, and democracy and how these are used to interrogate local concerns. The second section focuses on Cote d'Ivoire's *Zouglou* and *Coupé-décalé* musics, examining their engagement with the country's sociopolitical processes. New media technologies are noted as having facilitated its diffusion and established connections between Ivorians at home and the Ivorian diaspora in Western metropolises, most particularly Paris. The last section explores *Yabis* music in Nigeria. Largely attributed to the famous Nigerian musician Fela Anikulapo-Kuti, the section notes the music's enduring popularity and relevance as a means through which social and political criticism is still done. Its exploitation of the digital platforms has now facilitated its use by cellular phone companies, enabling its diffusion to millions, resulting in a powerful platform for social and political critique.

Urban Popular Music and the Mediation of the Public Sphere in Kenya

In 2002, two young Kenyan musicians, Joseph Ogidi (Gidi Gidi) and Julius Owino (Maji Maji), released a hugely popular single *Unbwogable.* The title of the song borrows from the English and Dholuo languages, with the English prefix "un-" joined with the Luo verb *bwogo* (to frighten) and the adjective "-able" added as a suffix. Distinctive in its grammatical transgression of the two languages, the song was, however, noted more for its deliberate transgression of official narratives of statecraft. The song's subtext contested notions of a harmonious, cohesive, and developing Kenyan nation. Instead, it expressed the angst and aggravation of two young men, partly blaming their broken dreams on politicized ethnicity, on their exploitation by the powerful, which had left them economically disenfranchised. *Unbwogable* soon morphed into a symbolic, almost parabolic, representation of Kenya's history of systemic political betrayal and social decay largely attributed to a rapacious political class. But the song also became a metaphor of triumph by the country's underclass. This metamorphosis was a result of various factors, not least the song's seeming contradictory appropriation by the political class. Having recognized its huge popularity, the National Rainbow Coalition (NARC), an alliance of fifteen opposition parties fighting to end Moi's twenty-four-year rule, appropriated the song—effectively making it their political anthem. *Unbwogable* suddenly became a veritable idiom of political resistance and freedom across the board, its spirit of defiance invested with national political agency. Yet the song's appropriation by the political class equally problematizes our understanding of popular music and more generally the "popular." For once the "popular" becomes part of official discourse, it may become a prop for statecraft—legitimizing the very idiosyncrasies it intended to ridicule.

Kenya's Political History and the Significance of the Realm of the "Popular"

Why is popular music so relevant to the discussion of the public sphere, and more specifically the political sphere, in Kenya? The practice or exercise of power in Kenya, as in many parts of Africa, cannot be confined to formal political structures. Informality has always been the continent's response to exclusion from structures of governance. Thus, it is partly in the informal arena, largely the realm of the "popular," that power is conceived, affirmed, and simultaneously contested by the political underclass.

Indeed, Kenya's silently turbulent recent political history is inextricably linked with the realm of the "popular." The country's first postindependence government under Kenyatta slowly muzzled the very freedom it fought for, as internal political competition grew within the nascent state. As a consequence, outside the institutional framework of state power, the government adopted various "justifying ideologies" of governance, reinvented various traditions, constructed myths, and employed useful imagery and metaphors as important instruments in legitimizing its rule (See for instance Atieno-Odhiambo 2002, 1987). Atieno-Odhiambo (1987, 179) writes about Kenyatta's construction of an "ideology of order," legitimated on the fact that there were various competing interests—religious, ethnic, and regional—which, unless checked, would impede the country's development. Through instruments of the state, such as the state-controlled media, Kenyatta brooked little opposition, so that at the time of his death in 1978, he was a hugely polarizing and unpopular leader. Broadly, state power rested with a select few from his extended family, and acolytes mainly from his Kikuyu tribe. Ethnicity had been heavily politicized to the extent that many saw the government as a Kikuyu government. Aware of the polarization and ethnic tensions, Kenyatta's successor Moi equally had reasons to institute an "ideology of order," which saw Kenya become a de jure one party state as he consolidated his power. Political repression continued, so that by the mid-1980s Kenya was effectively an artificial state; an unstable amalgamation of ethnic balkans controlled by the state through satellite ethnic barons created and sustained by the state mainly through clientalism and patronage. Political dissent inevitably began to fester.

Cultural production often intensifies at moments of political turmoil precisely because it is the condition of turmoil that invariably engender the production of the popular (Smith 1989). Because of the exclusionary tendencies of the state, the systematic disenfranchisement of millions of Kenyans, and the fact that public sites of communication were monopolized by the state, various sites of cultural production provided alternative spaces for political engagement and for the development of a counter-hegemonic discourse. Musicians, writers, and thespians captured how an oppositional cultural and political aesthetic flourished even amid the various constraints on not only the "expressible" but also the "thinkable,"

Popular music was especially engaged with the political sphere. Daniel Owino Misiani, a Luo musician of mixed Kenyan and Tanzanian heritage, represents a group of popular musicians who were at the forefront of articulating dissent through music from the 1970s through the 1990s. These musicians were explicit in their political intentions even though they mainly used allegory to indict both the Kenyatta and Moi regimes. Misiani's case is singularly important because, quite apart from the fact that until his death he was arguably one of the most popular *Benga*[1]musicians in Kenya, he was also Luo, a tribe that had considered itself systematically "persecuted" by both the Kenyatta and Moi administrations. Ogude (2006, 183) argues that in Misiani's songs, the Luo were a moral and a political category, a usable site of difference from which to critique Kenya's repressive polity.

Political Pluralism, New Media, and Urban Popular Music

As the media infrastructure developed in Kenya in the 1990s following the liberalization of the media sector, so did urban popular music, a nascent genre largely associated with the youth but which had gone largely unnoticed. By 2002, a vibrant, young, trendy, and experimental group of musicians, among them Gidi Gidi-Maji Maji, had emerged within this new context of political pluralism and media liberalization. Combining the already popular American hip-hop and rap genres with local styles and languages, this group of young musicians developed two distinctive musical styles, most notably *Genge* and *Boomba* (also known as *Kapuka*). Both were a complicated fusion of hip-hop, dance hall, and a motley of Kenyan musical styles. Key to their popularization were two production houses, Calif Records (based in Nairobi's Eastlands, a working-class neighborhood in the outskirts of the city) and Ogopa Djs. *Genge*, Kiswahili for "a mass of people," became emblematic of Nairobi's restive youth. Yet, unlike their predecessors, this music was defined by its focus on the banal, on the politics of the everyday, rather than the explicitly political. Romance, identity, and narratives around gender were this music's key features, although it was to be distinguished by its transnational character. Yet this is not to suggest that the music was necessarily apolitical. Instead, the said features functioned as tropes through which issues such as political exclusion, ethnicity, corruption, and various social problems were examined. Indeed, by celebrating Eastlands, a popular setting for many *Genge* songs, the place becomes a discursive symbol where power is registered and interrogated.

Political interventions were also made through very unique narratives. In a popular hit single, *Adhiambo Sianda*[2]—a musical collaboration between a Tanzanian group Deux Vultures and a popular Luo musician Dola Kabari,—a lady's voluptuousness and especially her big bottom provides the peg for the celebration of Kenya's ethnic diversity. Below is an excerpt.

> What makes Adhiambo look so good when she turns around
> Who can Bwogo me [Who can frighten me]
> *Wasichana wote wa Kenya duto beyo* [All Kenyan ladies are beautiful]

This song is not merely an affirmation or celebration of the beauty of the ladies (from Nairobi and Mombasa cities as well as Maasai and Nandi ethnic groups); it is also a significant political statement on some unifying notion of Kenyanness. This song is a very deliberate attempt at articulating, reflecting on, and depoliticizing ethnicity.

The song *Kenyan Boy, Kenyan Girl*[3] by Nazizi Hirji, stage named "Kenya's First Lady," also provides an interesting reading of a new form of political engagement. The song's storyline is about a girl who falls in love with a boy during a journey in one of Nairobi's Eastlands *matatus* (mini-buses). In the song, Eastlands is praised as are its inhabitants and, of course, its *matatus*. The song ingeniously interrogates notions of political and economic exclusion but equally stresses how a forgotten people have managed to survive this exclusion, even triumphing over it. The *matatu* in Kenya is more than just a means of transport. Over time, it has come to represent a culture of transgression, a statement of triumph over adversity (See Mbugua 2007). Using the *matatu* and Eastlands, largely representative of Nairobi's poor masses, the song celebrates their victory against their marginalization by the state. They are not victims but victors. The *matatu* is now the subject of various scholarly works. Atieno-Odhiambo (1987), for instance, has described the *matatu* as a "mini-republic." Indeed, at the heights of political repression during Moi's reign, it is in *matatus* that people discussed political issues, often disguised as rumors.

But focusing largely on the trope of romance has also meant experimenting with and stretching the bounds of the expressible. A number of musicians have been accused by the mainstream media and politicians of imitating American hip-hop music, glorifying and thus encouraging irresponsible youthful indulgence in sex, drugs, and crass commercialism. Hubert Nakitare (Nonini) released a succession of hugely controversial songs, including *Manzi wa Nairobi* (The girl from Nairobi), *Wee Kamu* (You come), *Keroro* (Alcohol), and *Hanyaring Game* (The dating game). The songs' explicit sexual lyrics caused widespread trepidation, with politicians, parents, teachers, and the clergy demanding that they be banned. Similar songs by artists such as Paul Nunda (Jua Cali), Peter Gatonye (PiliPili), and James Wathigo (JimW@t) among others elicited similar reception. It was evident the mainstream media was nervous about this newly found openness, especially to matters touching on sex and sexuality. Many musicians began to partly migrate to the "alternative" online platforms, although some had limited success.[4] Indeed, the web is a platform used by local musicians mostly for exposure especially to Kenya's diasporic communities rather than local audiences. But one also needs to recognize that this group is now becoming increasingly influential in shaping democratic processes in Kenya.[5] Thus, many established musicians now have pages on social networking sites, such as MySpace, and consistently post material on YouTube and various blogs. Because this virtual "networking" world is now increasingly influential in their "real" worlds, it has become a significant audience, shaping and reshaping the genre.

Significantly, these new platforms also provide a remarkable opportunity to musicians not only as spaces for exposure but an opportunity to visually express some of their encrypted lyrics through the use of promotional videos. Muhammed Swabiri, who fondly calls himself Red San, prominently features a Kenyan flag as a backdrop to his songs. Thus, for instance, in the single *Kenyan*,[6] partly sang in faux Jamaican patois, there are shots of Kenya's past and present leaders, as well as athletics and rugby sportsmen. But more significantly are shots of markers of affluence: big cars and designer wear, alongside handcarts and ordinary folk. While whipping up patriotic fervor with a popular Jamaican *ragga* chant of "Give them a signal," the contradictions that plague the Kenyan society are demonstrated in sharp relief.

This urban music's affinity to hip-hop music and its transnational character is also exemplified by Jackson Makini (CMB Prezzo), the self-proclaimed "King of Bling." A cult figure with the youth, Makini is known for his flamboyance: gold chains, expensive cars, and designer outfits. The mainstream press has often described him as the face of crass commercialism and of the youth imitating American hip-hop stars. But one may argue after Appadurai that as cultural forms move from the global mediascape into local ethnoscapes, their meanings are already altered in certain ways. Africa's conurbations are significant sites reflective of various forms of cultural hybridity (Appadurai 1991). They feature a cultural métissage definitive of postcolonial hybridity. They demonstrate neither a rapture with the past and its attendant cultural values nor a privileging of the present. The local is not dominant, but neither is the transnational or global, instead it is, to metaphorically use the words of Young (2001, 69), "a fragmented and hybridized language" emerging from a "conflictual cultural interaction," allowing for the creation of a new local identity that grafts the foreign into the local. Indeed, in Makini's case, although he is fond of the "*bling*" culture, he also sings in *sheng*, a creolized language associated with Nairobi's underclass.[7] Thus, one sees how these musicians mesh the local with the global and how fantasy and reality overlap in that continuous construction and reconstruction, especially of urban identities. Kenyan urban popular music is such a voice inhabiting, as it must, different worlds—but appropriation of the transnational is a fragile and complicated process, and therefore one necessarily fraught with contradictions. Consequently, while artistes such as Gidi Gidi-Maji Maji and D. O. Misiani sing from a site of difference, and thus betray elements of an ethnic cosmology of power, others such as Nazizi, Prezzo, Nonini, and Jua Cali may conceive of power differently in their construction of alternative realities—social or political. The consistency in this music is to be found in its inconsistency.

On the whole, one may argue that, with increasing exposure to new technologies and new media platforms, urban popular music will remain central to the mediation of the public sphere. Presently, because of problems of access and a developing media infrastructure, the role of new technology in this process, considering its potential, remains limited. It is important to note that in many developing economies such as Kenya, the exploitation of new media technologies still requires institutional support, which is problematic because by and large it also makes possible subservience either to big business or to the state, which leads to fears over their role as spaces for alternative discourses. But the ingenuity of Kenya's urban popular musicians demonstrates how improvisation can still provide spaces for subversion. In any case, it is improvisation that makes the popular realm popular.

New Media and Ivorian Music: The Representation of Postcolonial Relations with France in *Zouglou* and *Coupé-Décalé*

In Côte d'Ivoire, popular music genres such as reggae and *zouglou* have played a critical role at significant turning points in recent history, and have served as a domain for articulation of ideas about politicians, corruption, citizenship, national history, and identity. *Zouglou* emerged in the 1990s in the context of the student demonstrations for political liberalization and, along with reggae, served as a platform for criticizing prevailing social and political conditions. Ivorian popular music has consequently been associated with the return to multiparty politics under Houphouët-Boigny in 1990 and the overthrow of Bédié by General Gueï in 1999. During the civil war that broke out in September 2002, many well-known *zouglou* artists have commented on the crisis in "patriotic songs," which can be interpreted as supporting Gbagbo's government and the Alliance of Young Patriots led by Goudé. In recent years, *coupé-décalé*, a new, ostensibly apolitical, musical genre has emerged, and its popularity has been giving solid competition to *zouglou* as the most celebrated Ivorian music. However, *coupé-décalé* is effectively also engaging with a sociopolitical project, as its imagery envisions a new postcolonial relationship with France. As such, it provokes a break with the tradition of social criticism represented by *zouglou*, where *zouglou* artists, presenting themselves as the "voice of the voiceless" depict themselves as social underdogs at the receiving end of an unjust and unequal postcolonial relationship. Born in the nightclubs of the Ivorian diaspora in London and Paris, *coupé-décalé* is a transnational musical style *par excellence*. New media technologies have played a notable part in the diffusion of *coupé-décalé*, and in connecting Ivorians at home with those living abroad.

Zouglou emerged in the early 1990s out of a movement for democratization that was led by students at the University of Abidjan, and presented itself as a genre of marginalized youth. Since *zouglou* emerged in the context of political protests, zouglou's many songs about the adverse living conditions of students were associated with criticism of the failing economy and the Houphouët-Boigny government. Zouglou emerged in an "atmosphere of general struggle" and animated the meetings of FESCI, the Students' and Schools' Federation of Côte d'Ivoire (Thiemélé 2003, 71). The deterioration of the standard of living certainly affected the population at large, and widespread public unrest ensued, including strikes in the public sector and security forces. Popular music, especially reggae and *zouglou*, turned into "key channels for raising political awareness and mobilization" (Akindès 2002, 86). As a new, urban musical form, *zouglou* distinguishes itself through its use of *nouchi*, the French street-slang in Abidjan, and its use of very direct, outspoken texts, rather than of subtle, coded messages. This may partly be due to *zouglou*'s birth in the context of the liberalization of the press, and a perception that a frank discussion of the country's problem was long overdue.

Zouglou, as a musical genre that has functioned as political activism since its inception, directed its social criticism and condemnation of injustice not just at Ivorian society but also toward the former colonial power France. The song *Sans papiers* (illegal migrants), by the *zouglou* group Les Salopards, denounces the unequal conditions of travel between France and Côte d'Ivoire, since, according to the song, the French can travel without impediments to Côte d'Ivoire as *coopérants* or as tourists, yet Ivorians are required to present their visa and residence permits upon arrival in France.[8] The song additionally comments that relations between blacks and whites have always been hypocritical, that understanding between blacks and whites has never existed, since even though the earth and heavens belong to God, there are countries that are forbidden to blacks.[9] Even though the tune is rather slow and mournful than is typical of *zouglou*, the straightforward and outspoken manner of delivering sharp criticism is characteristic of *zouglou*'s aesthetic.

Another characteristic feature of *zouglou* is the humorous and often story-like texts. The song *Un gaou à Paris* (a fool in Paris) by the *zouglou* group Magic System speaks of the difficulties of an Ivorian in France in the joking manner that *zouglou* has become so famous for. This song is less somber, and entails a rather satirical depiction of the life of Ivorians in France. To illustrate, the vocalist sings that everything in France is machinated, even to buy food, one speaks to machines; in the video, the song's protagonist suffers in Paris—walking through the drive-through part of a fast-food restaurant to order a meal.[10] To quote the song: "to get papers, one suffers, oh. Upon arrival over there, one suffers. Suffering upon suffering, oh."[11] In the France of *Un gaou à Paris*, it is so cold that smoke comes out of one's mouth even when not smoking. However,

France is also socially cold; it is the country where no one says hello, everyone is in a hurry, everyone fends for himself, no one lends money, and is a place where you just have to try to get by. Even if you left Abidjan with lots of money, upon arrival in France, your money is worth little.[12] *Zouglou* stars Petit Yodé and L'Enfant Siro sing, in their song *Bengué*: "They called their country France, we blacks, as soon as we heard 'France,' we thought directly about the Franc. But when they say France, the Franc is for the whites and the suffering (souffrance) is for the black" (Yéré 2006, 83). Musicians representing themselves as "the voice of the people," according to *zouglou* star Petit Yodé, condemn the unequal postcolonial relationship with France, citing the racism, inequality, and lack of hospitality in France.

However, the new musical style *coupé-décalé* portrays a radically different existence for young Ivorians in France. In *coupé-décalé*, the trajectory is the opposite of that portrayed in *Un gaou à Paris*: young Ivorians arrive in Paris and, soon after, leave France pockets bursting with money to go back home to celebrate their success abroad. Thus, *coupé-décalé* as a genre celebrates the exploits of young Ivorians having made a fortune in France. Rather than suffering from the injustice of the postcolonial relationship, in *coupé-décalé*, Ivorians themselves celebrate for mastering both environments. Therefore, Barber and Furniss' suggestion that new genres often imply "new ways of looking at the world" (2006, 10) is demonstrated in the new sensibilities associated with the development of *coupé-décalé*.

A major element that contributes to the success of this new musical style is the idea of travel and movement. As noted by Kohlhagen, while the protagonists of *coupé-décalé* praise themselves of their access to consumer society, the means of reaching that point is travel to Europe (Kohlhagen 2005, 98). Kohlhagen also points out that even more than its texts, it is the images that *coupé-décalé* conveys that have contributed to its success. The music videos of these songs are filmed in European capitals where the Ivorian stars dance on the grandest avenues and in front of the most prestigious historic monuments. For example, in the video to the song *Sagacité* (sagacity), the group Jet-7 strolls in the designer boutiques of the Champs-Élysées, distributing €100 bills to passersby and teaching Europeans to dance *coupé-décalé* in front of the Eiffel Tower (Kohlhagen 2005, 96). In this song, between exclamations of "*coupé*" and "*décalé*," the names of towns are inserted: London, Brussels, Paris, Geneva, but also Abidjan, Daloa, and so on—signifying a movement back and forth between the continents. The axis Abidjan-Paris is very present in the video clips. In *Héros national* (national hero), for example, Douk Saga rolls simultaneously in front of La Defense (Paris) and the district of the Plateau (Abidjan) in his convertible, the urban backgrounds merging (Kohlhagen 2005, 103). *Coupé-décalé* therefore does not present itself as the music of exiles having succeeded in the adaptation to a new social environment, but rather as the music of travelers knowing how to benefit from that foreign environment (Kohlhagen 2005, 102). The stay in Europe is not a definitive one, and the Europe of *coupé-décalé* is an African Europe (Kohlhagen 2005, 104), in which the local European population is relegated to the background.

Therefore, in *coupé-décalé*, a new relationship between France and Côte d'Ivoire is imagined and depicted, one that is at the polar opposite of the image portrayed in *zouglou*. In *zouglou*, where Ivorians have to cope with the climatic and social harshness of Europe, in *coupé-décalé*, they profit from their residence in Europe. While in *Un gaou à Paris*, the Ivorian is the fool, in *coupé-décalé*, it is Europeans who are fooled due to the music's suspected connection with illegal practices of accumulation. Though originating in the capitals of Europe, *coupé-décalé* is very successful in Abidjan, and its success is expanding beyond Côte d'Ivoire into the other African countries, such as Cameroon and the two Congos. It is a musical style that is emblematic of modern technologies, both of movement and of communication, connecting both Ivorians at home and overseas. In fact, both *coupé-décalé* and *zouglou* are diffused on Internet sites such as YouTube, and on the online radio services of Abidjan-based stations such as Nostalgie and Radio Jam, as well as the stations of Ivorian diaspora based in Europe and in the United States.

Radio is the most widespread source of information and entertainment in Côte d'Ivoire, and it is also the Ivorian media that is the most accessible one outside Côte d'Ivoire, since several Ivorian radio stations broadcast online. On these radio stations, there are wish list programs, where listeners can phone in or e-mail in from Côte d'Ivoire and from overseas and request songs for friends and relatives living either in Côte d'Ivoire or abroad. A listener request e-mailed to an Ivorian radio station can thus be played for listeners on the radio in Côte d'Ivoire, while those in the diaspora listen to the same request on line. Also, many of the video clips of the latest hits of *zouglou* and *coupé-décalé* can be viewed on YouTube and on Ivorian TV. In fact, music TV programs aired on Radio Télévision Ivorienne are recorded and sold in Ivorian shops in Europe on DVD. These new media technologies connect Ivorian communities at home and in the diaspora, while young Ivorians at home and abroad negotiate new (musical and symbolic) relationships with France.

"Digital Nigeria": *Yabis* Music and the Political Public Sphere

Yabis is the noun of the Pidgin English word "*Yab*" (which literarily means to make fun of a person, a group of people, or a thing). The etymology of the word "*Yab*" is difficult to trace. But, in terms of its sound, two English words that come to mind are yap and yak. While the former is defined as talking noisily without saying anything very important or serious, the latter means talking continuously about things that are not very serious. However, when closely compared with the word "chatter," which is a synonym of yak, then "*Yab*" may have a closer affinity with yak than yap.

Yabis, in this study, is defined as a biting satirical song deliberately composed with the aim of correcting an atrocity, a misdemeanor, or a sacrilege committed by either an individual or a corporate body within a particular society. When viewed in line with some forms of performances in other creative arts disciplines such as theatre arts, film, and dance, one may convincingly assert that *Yabis* music, like *Yabis* comedies, exposé plays, and some other forms of satirical songs rendered in various traditional African ethnic languages, are effective means of correcting vices and misdemeanors in a society.

One Nigerian musician who pioneered and saw to the propagation of *Yabis* music in Nigeria was the late Fela Anikulapo-Kuti, formerly known as Fela Ransome-Kuti. Fela was the second son of the legendary Kuti family of Rev. and Mrs. Ransome-Kuti in Abeokuta. On return from his American tour in 1970, Fela came up with a different idea and concept about African culture and traditions, most especially in the area of African music and politics. But the turning point in Fela's life came after meeting Sandra Smith, a member of the Black Panther Party in the United States. She exposed Fela to the likes of Marcus Garvey and Malcolm X, so on his return to Nigeria, Fela decided, among other things, to change the name of his band from *Koola Lobitos* to *African 70.* Second, he abandoned his former highlife-jazz style for what he called "Afrobeat." Also, his themes now dealt with the daily social struggles rather than his usual emphasis on love and romance. Thematically, *Yabis* music was to become similar to most Black or Afro-American music, such as Blues, Soul, Spiritual, Reggae, Calypso, and Hip-Hop, which tend to feed on a narrative that privileges the struggle for liberty.

Yabis Music and Pidgin English—A Synergic Paradigm

For any language to survive in a heterogeneous nation like Nigeria, it must be accessible across social classes—which explains the conviviality of Pidgin English in Nigeria. One advantage that Pidgin English has over the English language is the retention of the sounding qualifiers. Most African indigenous languages have a different concept for qualifiers compared to English. While English adverbs or adjectives would qualify a noun or verb, in most African languages, if it doesn't sound it, it doesn't describe it, and if it doesn't describe it, it doesn't qualify it. In other words, qualifiers must also be descriptive (Olatunji 2001, 45).

While an English speaker would say: "He gave her a 'dirty' slap," a speaker of Pidgin would say: "He gi am slap for face, '*gbosa*.'" So, the use of Pidgin English for *Yabis* music creates ample opportunities for an extensive integration of onomatopoeias. Consequently, there is usually an avalanche of several onomatopoeic words ingrained in the lyrics of *Yabis* musicians. For example, in an effort to vividly describe the barbaric action of the military men who forcefully broke into and plundered his former "Kalakuta Republic" house, in his music titled *Unknown Soldier*, Fela sang:

> ...Then suddenly, suddenly...
> *Jagbajagba, Jagbajagba, jagba*
> *Jugbujugbu, Jugbujugbu, jugbu,*
> *Jigbijigbi, Jigbijigbi, jigbi*
> *Jegbejegbe, Jegbejegbe, jegbe...*

All these are nothing but nonsensical syllables adopted as sound qualifiers to portray the moments of looting, plundering, breaking, and entry that took place as well as the attendant commotion, wailing, sorrow, and brouhaha that ensued afterward (Olatunji 2007, 6).

In the same vein, Eedris Abdulkareem, a Nigerian rap music star, employs two onomatopoeic words, *jagajaga* and *gbosa-gbosa* in his music titled *Jagajaga*. Through the former, he depicts the tragic situation similar to Hobbe's "state of nature" in Nigeria and by the latter he represents the sound of sporadic gunshots of the armed robbery gangs operating in every nook and cranny of the country. An excerpt of the song goes thus:

> Nigeria jagajaga,
> Everything scatter scatter,
> Poor man dey suffer suffer,
> Gbosa! Gbosa!!
> The Roles of Yabis Music

In line with its satirical nature, sometimes the performance of *Yabis* music is preceded or interpolated with either a speech monologue by the band leader or a dialogue by the leader and the rest of the members of the band. For example, Fela usually begins or interpolates his *Yabis* songs with a dialogue between him and members of his African 70 during recording sessions, or with the audience joining during his live shows. The dialogue mostly goes thus:

Fela: Everybody, say ye-ye
All: Ye-ye

Interestingly, when the two syllables in *ye-ye* are joined to form a word and tone-marked with low and high speech tones (\ and /), it forms the word *Yè'yé'*, which means satire in Yoruba language. At other times, Fela's dialogue takes this form:

Fela: Make I yab dem?
All: Yab dem.

Similarly, Eedris Abdulkareem starts his song titled *Letter to Mr. President* with a monologue: "Turn dat microphone on, make I yarn some reality..." However, African China, in his record titled *Mr. President*, started with a cockcrow sound, perhaps to signify an important early morning message from him to his fellow countrymen. Despite being largely satirical in both its concept and content, *Yabis* music plays other roles.

Yabis motivates and inspires listeners to stand up against various forms of injustice. But such action is not limited to targeting the political class. It is as much targeted at ordinary folk. Indeed, as Oyewo (2003, 59) acknowledges, oppression and its odious accomplice, tyranny, is not unique to rulers alone. Precisely because of Nigeria's political history, one dominated by military rule, the relationship that successive governments had with the public is employed as a trope in *Yabis* music to map and critique contemporary social and political experiences.

Yet in *Yabis* music we also note the realm of the "popular" as notoriously complex and necessarily inhabited by many voices. *Yabis* music is a "free zone" for both the government and the masses. On one hand, *Yabis* music has been used in diverse ways to critique the country's ruling class, but, on the other hand, the ruling class has also found *Yabis* to be a veritable site through which it spreads its propaganda. Quite often, several government agencies, such as the ministries and parastatals of both the federal and state governments, utilize this "zone" to advertise their various programs to Nigerians. For instance, awareness information jingles on HIV/AIDS, child labor, and women liberation/ emancipation feature prominently on both government and privately owned radio and TV. Here is an excerpt from the U.S. Agency for International Development sponsored jingle on HIV/AIDS:

...One day one guy, one day one guy...
One day one girl, one day one girl,
Many partners *plenti wahala*
One partner less palaver...

Also, during the 2007 electioneering campaign in Nigeria, most political parties had some of their jingles rendered as *Yabis*. A common example was the Peoples Democratic Party's *di koko* jingle, which became very popular among the youth in the country. *Di koko* is best translated as "the only original," which in this context means that every other political party at that time was fake or inferior. The song that was originally composed by D'bang, a Nigerian Hip-Hop music exponent, was not meant for any political party's campaign rally. But one adapted and modified version used as a campaign rally jingle goes thus:

> Make I tell you *di koko o, ah!* Make I tell you *di koko*
> Make I tell you the *di koko,* if you want to know *di koko*...
> *PDP na di koko, ah! PDP na di koko*...

Yabis music, like any other brand of music or performing arts, portrays life in its base society—because artists create their works from the raw materials supplied by events in the society. For example, Fela's first hit record was titled *Jeun ko o ku* (A Glutton), where he illustrated the wretchedness of people who, in the Nigerian perspective, use their "ten fingers" to eat. This metaphorically refers to those who squander all their income today, not saving for tomorrow. Ever since, there has been an avalanche of *Yabis* songs both by Fela and other Nigerian popular musicians in this direction.[13]

Yabis music must also be located within a postcolonial paradigm in which European cultural values are constantly challenged. Indeed, the political culture it envisions is not necessarily European. Similarly, it also challenges a range of foreign (including Christian and Arabic) cultural values as symbolized by people's names and dressing. It attempts to promote what it considers local.[14]

The Impact of Technology on Popular Music in Nigeria

From the 1990s, technology has had an impact on *Yabis*, alongside other contemporary Nigerian popular music. Our discourse in this regard is centered on three major technological media: radio and TV, GSM cellular phones, and the Internet.

The advent of private enterprise in the 1990s brought the emergence of privately owned FM radio stations and TV stations in Nigeria. Consequently, this provided the much-needed avenue for the growth of urban popular music in the country. In 1992, Chief Raymond Dokpesi established Ray Power 100.5 FM as a wing of his DAAR Communication outfit. It was the first station to start the twenty-four-hour broadcast tradition in Nigeria. This, indeed, created more space for songs by Nigerian popular music artistes. Some years later, other radio stations such as Star FM 105.5 and Brilla FM emerged—resulting in competition. In 1995, DAAR Communication also established the first private

TV station, African Independent TV. Several other stations sprang up soon after.[15] Also, like its radio wing, AIT started a twenty-four-hour broadcast, then unprecedented in Nigeria, providing multiple avenues for the promotion of *Yabis* music.

It is also worthy of note that today favorite tracks of Nigerian popular musicians' albums are used as signature tunes by both the radio and TV stations for their programs. Most of these tunes are also used as jingles for various adverts both by government departments and parastatals or producers of goods and services.

At the turn of this millennium, with the advent of satellite broadcasting in Africa, *Yabis* music found a platform to a global audience. Today, many Nigerian TV stations—both private and state-owned—are connected to satellite.[16] Consequently, Nigerian music and film industries, together with their practitioners, have gained recognition and popularity throughout Africa and the rest of the world.

The civilian government that came to power in 1999 also allowed for the influx of both the GSM service providers and cellular phone market in the country. Today, there is an avalanche of these telecommunications service providers in the country (among them MTN, Starcoms, Globalcom, Celtel, Intercellular, Multilink, and Oduatel). There is also a big market in the country for cell phone manufacturers such as Nokia, Sagem, Motorola, Sony Erikson, and a host of others. Cell phone users can now enjoy such services as tuning to radio stations, recording and playing back music, or speech data, as well as several Internet services.

However, two of these facilities are germane to the current study. These include the ring tones and caller-tone facilities. Today, most GSM service providers have several tunes of many Nigerian popular musicians stored in their tone banks, and customers can download either ring tones or caller tones from these tone banks. Mostly, musicians whose songs have been stored and the tracks stored are advertised both on radio and TV. Customers are also given some numbers to use as dial codes before accessing these tunes.

The Internet is the meeting point between the radio/TV and the GSM facilities. All major radio/TV houses as well as the GSM service providers and cell phone manufacturers have their Internet Web sites. Indeed, the storage and access of tones could not have been possible without the Internet. With the expansion of digital technology and its uptake in Nigeria through the rolling out of broadband, *Yabis* music, like other cultural forms will gain from yet another platform that will enable it increased access to global audiences. Similarly, because of the relative freedom of the Internet from various legislative constraints vis-à-vis the more generic broadcast media, *Yabis* music will most likely exploit this new space as another powerful site for social and political critique.

Conclusion

Certain trends emerge from our analysis of the engagement of popular music with new media from the 1990s onward. This period coincides not only with the expansion of new media in Africa but also with radical political transformation in the continent. This has led to a notable liberalization of the media, which has since facilitated new ways in which popular cultural forms dialogue with transnational discourses, especially in so far as they engage with political processes and identity politics. The reception, however, remains varied, yet it merely emphasizes the equivocality of culture. In Kenya, urban popular musicians have been accused by the mainstream media and politicians of glorifying and encouraging irresponsible youthful indulgence in sex, drugs, and crass commercialism. In Côte d'Ivoire, a similar charge is made against *coupé-décalé*, a genre perceived as encouraging ostentatious behavior. Yet the argument is far more complicated. We have noted how fantasy and reality overlap, which is, in part, a survival strategy in Africa's challenging urban environments. In addition, we have argued that these groups engage with politics in new ways. Thus, what may be dismissed as mere entertainment may not in fact be apolitical.

In Côte d'Ivoire, Kenya, and Nigeria, popular music continues to serve as an alternative space for political engagement. Yet we have argued that one must recognize the complex nature of the "popular," especially the fact that it is not exclusively a province of the political underclass—neither is it necessarily counter-hegemonic.

In both Kenya and Côte d'Ivoire, popular music also plays a role in the popular definition and conceptualization of Kenyanness and Ivoirité. This becomes especially apparent in the music videos now posted on sites such as YouTube, which itself is a transnational space. Thus Muhammad Swabiri uses a Kenyan flag as the backdrop of his videos, just as the video of Ivorian patriotic songs are filmed through a lens in the shape of the outline of the Ivorian state. In both settings, transnational images shape popular music's reengagement with national identity. Thus, the single *Kenyan* is sung in faux Jamaican patois, while Ivorian singer Gadji Celi sings patriotic songs such as *Ne touchez pas a mon pays* and *La mère patrie* to Congolese rhythms. New media have facilitated interaction with the transnational and thus have been crucial in rethinking the national anew and at the same time linking diasporic communities with those in the home country. The new media have therefore been influential in circulating this music and contributing to the global flow of styles, images, and narratives that shape various sociopolitical processes in Kenya, Côte d'Ivoire, and Nigeria.

Notes

1. *Benga* is a music style typically associated with the Luo from western Kenya.
2. Zogoklan 2007. *Adhiambo Sianda* [online] http://www.youtube.com/watch?v=_OZLOzW5ntc (accessed July 5, 2008).
3. hba84 2007. *Kenyan boy Kenyan girl-Nazizi feat. Wyre* [online]. http://www.youtube.com/watch?v=B6xGoWTGjpw (accessed July 5, 2008).
4. Internet usage, more recent usage via cell phones, has increased steadily in Kenya since the late 1990s. Conservative estimates put user numbers at 8 percent of the population. See. http://www.internetworldstats.com/af/ke.htm (accessed July 5, 2008).
5. Conservative estimates put this community's contribution to Kenya's economy at a whopping Ksh 53 billion (US$ 7.5 billion) annually.
6. Anelka1030. 2007. *Proud to be Kenyan-Redsan* [online] http://www.youtube.com/watch?v=M4ufvqH4e0s&mode=related&search= (accessed July 5, 2008).
7. For a detailed reading on *Sheng*, see Githiora, C. 2002. "Sheng: Peer Language, Swahili Dialect or Emerging Creole," *Journal of African Cultural Studies* 15(2): 159–181.
8. Original song text: *Quand ils viennent chez nous on dit qu'ils sont coopérants. Quand ils viennent chez nous, on dit ils sont des touristes. On va aller chez eux, on dit présente tes visa, on arrive la bas, on nous dit carte de séjour.* All translations by the author.
9. *Entre les blacks et les whites il y a des relations hypocrites. Entre les blacks et les whites l'entend n'a jamais existée. Oh que j'ai si mal de voir mon peuple humilie.... C'est à Dieu qu'appartiennent le ciel et la terre. Pour des problèmes de papier, il y a des terres qui nous sont interdites.*
10. *Là-bas tout est machine, oh. Même pour payer manger, oh, tu parles avec machine oh.*
11. *Pour avoir des papiers on souffre oh. Arrivé là-bas on souffre oh. Souffrance dans la souffrance oh.*
12. *C'est quel pays où y a pas bonjour oh. Tout le monde est pressé oh. Chacun dans son chacun oh. A Paris ya pas mon frère donne-moi crédit oh. On n'a qu'à se débrouiller oh.... On quitte chez nous avec beaucoup d'argent oh, Arrivé là-bas ça devient petit oh.*
13. Examples include Fela's *Water No Get Enemy*, *Trouble Sleep Yanga Wake am*, *Gentleman*, *Suegbe Na Pako*, *Shakara*, *Lady*, and *Original Suffer Head*; Eedris Abdulkareem's *Mr. Lecturer* and *Chop Life*; Lagbaja's *No do Gra-gra for Me* and *Simple Yes or No*; African China's *No Condition is Permanent*, *Men Wey Sabi* and *Last Year*; and Femi Kuti's *Victim of Life*, *Sorry Sorry*, and *Shotan*.
14. Examples of compositions that demonstrate this confrontation include Fela's *Colomentality*, *Beast of No Nation*, *Teacher No Teach Me Nonsense*, *Yellow Fever*, *Original Suffer head* and *Suffering and Smiling*; Femi Kuti's *Wonder Wonder*; and Lagbaja's *Afro-Calypso*.

15. These included Murhi International Television, Minaj Broadcasting International, Channels Television, Galaxy Television, DBN (Sports) TV, and Silverbird TV.
16 These include AIT, Gateway Television (GTV), Nigerian Television Authority (NTA), Silverbird, DBN and Channels.

References

Adedeji, F. and M. Olatunji. 2006. Reassessing the Significance of the Performing Arts in Our Contemporary World. Paper Presented at the 3rd Faculty of Arts Conference, June 12–15, Obafemi Awolowo University, Ile-Ife, Nigeria.

Afolabi, J. A. 2004. "Harnessing the Potential of the Performing Arts for National Development in the Era of Globalization." In *Locating the Local in the Global: Voices on a Globalized Nigeria*, ed. S. Akinrinade, M. Kolawole, I. Mojola, and D. Ogungbile, 149–158. Ile-Ife: Obafemi Awolowo University Press.

Akindès, S. 2002. "Playing it 'Loud and Straight': Reggae, *Zouglou*, *Mapouka* and Youth Insubordination in Côte d'Ivoire." In *Playing with Identities in Contemporary Music in Africa*, ed. M. Palmberg and A. Kirkegaard. Uppsala: Nordiska Afrikainstitutet.

Appadurai, A. 1991. "Global Ethnoscapes: Notes and Queries for a Transnational Anthropology." In *Recapturing Anthropology*, ed. R. G. Fox, 191–210. R. Santa Fe, NM: School of American Research Press.

Atieno-Odhiambo, E.S. 1987. "Democracy and the Ideology of Order in Kenya." In *The Political Economy of Kenya*, ed. M. Schatzberg, 177–201. New York: Praeger.

———. 2002. "Hegemonic Enterprises and Instrumentalities of Survival: Ethnicity and Democracy in Kenya." *African Studies* http://www.informaworld.com/smpp/title~content=t713405746~db=all~tab=-issueslist~branches=61—v6161 (2): 223–249.

Barber, K. 1997. *Readings in African Popular Culture*. London: The International African Institute in association with Indiana University Press.

Barber, K. and G. Furniss. 2006. "African Language Writing." *Research in African Literatures* 37(3): 1–14.

Kohlhagen, D. 2005. *Frime, Escroquerie et Cosmopolitisme: Le Succès du "coupé-décalé" en Afrique et Ailleurs.* In *Politique Africaine* 100: 92–105.

Thiemélé, R. 2003. *L'Ivoirité Entre Culture et Politique.* Paris: L'Harmattan.

Mbembe, A. and S. Nuttall, eds. 2004. *Johannesburg: The Elusive Metropolis.* Durham: Duke University Press.

Mbugua wa Mungai. 2004. Wahome Mutahi and Kenyan Popular Culture. Unpublished.

Nyairo, J. and J. Ogude. 2005. "Popular Music, Popular Politics: *Unbwogable* and the Idioms of Freedom in Kenyan Popular Music." *African Affairs* 104(415): 225–249.

Ogude, J. 2006. "The Cat that Ended up Eating the Homestead Chicken: Murder, Memory and Fabulization in D.O. Misiani's Dissident Music." In *Urban Legends, Colonial Myths: Popular Culture and Literature in East Africa,* ed. J. Ogude and J. Nyairo. New Jersey: Africa World Press and New Africa Books.

Ogude, J. and J. Nyairo, eds. 2006. *Urban Legends, Colonial Myths: Popular Culture and Literature in East Africa.* New Jersey: Africa World Press and New Africa Books.

Olatunji, M. O. 2001. "The Use of Pidgin English in Contemporary Nigerian Music: A New Approach Towards National Identity." *Humanities Review Journal* 1(1): 41–46.

Oyewo, G. A. 2003. "The Creative Artist and the Aesthetics of Ethos: A Study of Afolabi's 'To the Stars.'" *Nigerian Music Review* (4): 55–64.

Schatzberg, M. 2001. *Political Legitimacy in Middle Africa: Father, Family, Food.* Bloomington and Indianapolis: Indiana University Press.

Scott, J. 1990. *Domination and the Arts of Resistance: Hidden Transcripts.* New Haven and London: Yale University Press.

Smith, A. 1989. *East African Writing in English.* London: Macmillan.

Tudesq, A. 2004. "*La Télévision en* Côte d'Ivoire." In *African Media Cultures: Transdisciplinary Perspectives / Cultures de Médias en Afrique—Perspectives Transdisciplinaires,* ed. R. M. Beck and F. Wittmann, 241–262. Köln: Köppe Verlag.

Yéré, H. 2006. *The Cultural Biography of a Generation: Zouglou Music and Youth in Côte d'Ivoire, 1965–1999.* MA Thesis, Centre for African Studies, Universität Basel.

Young, R. J. C. 2001. *Postcolonialism: An Historical Introduction.* London: Blackwell.

CHAPTER 13

News Media Use of ICTs amidst War, Violence, and Political Turmoil in the Central African Great Lakes

Marie-Soleil Frère

For the countries of Central Africa, the decade of 1993–2003 was one of war, violence, and political turmoil. The genocide in Rwanda, two wars in the Democratic Republic of Congo (DRC) involving up to nine countries in the region, and ten years of civil war and massacres in Burundi turned the whole area into a battlefield, causing approximately five million deaths (Zacharie and Janne d'Othée 2004). The media and journalistic practices were deeply affected by this conflict that impacted upon all spheres: political, economic, and social (Frère 2007). Conflicts drive protagonists to control the media and to restrict access to information. Meanwhile, they make journalists unsafe, impoverish the population, and prevent audiences from buying newspapers, as potential advertisers vanish. Often, during conflicts, journalistic ethics suffer as propaganda thrives and drives out balanced news.

Nevertheless, during that same period, the journalists of DRC, Rwanda, and Burundi also experienced changes in their professional practices, thanks to new information and communication technologies (ICTs), which have had a growing impact from 2000 onward. This chapter aims to describe how ICTs, mainly the Internet and the cell phone, helped develop a better approach and understanding by local journalists of the complex situation in their countries. ICTs therefore transformed professional capacities and influenced the peace process, facilitating peace in the Great Lakes region.

Many theoretical approaches, such as "peace journalism" (Lynch and McGoldrick 2001), *conflict sensitive journalism* (Howard 2003), or journalists as "mediators" in conflicts (Baumann & Siebert 2000), emerged in this period; most of them put forward the basic principles that should guide any reporter covering conflict and peace-building issues—but none of them have underlined the impact of the recent technical evolutions on professional behaviors and contents. On the other hand, there have been attempts to show how ICTs are impacting on journalists' daily work in Africa (Berger 2005; Institut Panos Paris 1998), but none focused on how this might have contributed to a more balanced and responsible journalism in conflict situations. This chapter tries to show that the use of ICTs by the media has probably played a more positive part in the political and social evolution in the Great Lakes region than the promotion of appropriate professional principles on a theoretical level. Nevertheless, if they help give more voice to the usually silent local population, ICTs also raise some questions about their possible misuse as obstacles to peace-building in very fragile states.

New Technologies, New Opportunities for Great Lakes Journalists

The regional war in the Great Lakes region did not prevent the growth of the use of ICTs.[1] From 1999 to 2004, when the area was experiencing a regional war, cell phone use had grown by 162.9 percent in Burundi, 175.4 percent in DRC and 66.0 percent in Rwanda, reaching, in 2004 a ratio of 1.42 subscribers for 100 inhabitants in Burundi, 3.49 in DRC, and 1.64 in Rwanda.[2] During the same period, Internet subscribers had grown by 41.1 percent in Burundi, 64.2 percent in DRC and 50 percent in Rwanda, reaching a ratio of 0.35 Internet users for 100 inhabitants in Burundi, 0.09 in DRC, and 0.45 in Rwanda.[3] This evolution occurred as the main peace agreements were signed in 2000 for Burundi (transition period until 2005) and 2003 for DRC (transition period ending with the 2006 general election). The journalists in the region, who had previously been accused of promoting "hatred media" (the Burundian press between 1993 and 2006; the RTLM—Radio Télévision des Mille Collines—in Rwanda in 1994, the "anti-Tutsi" press in Kinshasa in 1998), were among the "happy few" that got access to those new technologies and benefited rapidly from their use.

ICTs brought significant changes in six areas of journalistic practice in the Great Lakes region: news-gathering, producing, distributing, and archiving, as well as journalists' protection and training opportunities.

These changes are similar to the general evolution of the media in many other countries, but they might have had a special impact in the very peculiar context of the war situation in Central Africa. ICTs appeared at the very end of the 1990s in an environment with two main characteristics: a financially very weak media (with limited audience due to difficult financial sustainability and hardly any capacity for any investment) and, paradoxically, a politically powerful media in a very fragile and unstable political and institutional context, where the means of communication proved they could be war weapons (Frère 2007, 11).

News-gathering: Multiple and People-based Sources versus War Propaganda

The first major change brought about by ICTs for the media operating in these unstable states was the unprecedented number of sources of information, both through the Internet and cell phones. The war in Central Africa had for a long time resulted in propagandist media, partly due to the lack of capacities to cross-check sources or to get access to other perspectives or points of view. Rwanda's RTLM radio was so successful in preparing the minds for the Tutsi genocide and helping its execution because, besides the government-controlled Rwandese Information Office (ORINFOR) radio station, the Rwandese population did not get access to any other source of information. During the whole genocide (from April 7 to July 4, 1994), the main telephone lines were not operating and the Rwandese population did not get any news other than from the propaganda broadcast by RTLM and ORINFOR (Chrétien 1995, 284–286).

Before the end of the 1990s, if Congolese or Burundian journalists wanted to listen to alternative points of view, they could only listen to international radio stations such as Radio France Internationale (RFI), BBC, or VOA, which broadcast only in a few African local languages.[4] They had no access to the news produced by their colleagues in the neighboring countries. When DRC was split and occupied by several armed groups, journalists based in Kinshasa could not know exactly what was going on in Eastern Congo; to get information about the situation in Bukavu or Kisangani, they had to listen to international radio stations.

The Internet and cell phones profoundly changed that situation. Colette Braeckman, a Belgian journalist who has covered the Great Lakes for decades, is convinced that circulation of information via the cell phone was among the factors that put an end to the war in DRC; she argues that if cell phones had been operating in Rwanda in 1994, the genocide might have been avoided and the international community would have been forced to confront the situation and intervene.[5]

The Internet and cell phones gave Rwandese, Congolese, and Burundian journalists an unprecedented opportunity to get access to multiple sources of information about what was going on in their countries.

In most press enterprises in Africa, documentation is scarce. Subscription to big press agencies is too expensive for media that are often facing problems to pay the wages of their own employees. Through the net, small newspapers and radio stations all over the huge Congolese territory could get information freely from mainstream or alternative media Web sites. For example, it is thanks to the Internet that the Report of the UN Experts Panel on the Illegal Exploitation of Natural Resources in DRC or the Dar es Salaam Declaration adopted by eleven heads of state at the end of the war in 2004 became available to all African journalists, including the ones from the countries directly involved. Such documents would have been totally unknown by local media before the development of the Internet. From these documents the local journalist could know the exact commitments made by the heads of state, enabling them to monitor if they would act according to what they had officially agreed to.[6]

Circulating official documents, books, reports or any kind of printed information had been almost impossible for years in a country like the DRC, with hardly any tarmac roads or mail delivery facilities. More than eighty publications and almost 170 radio stations in the provinces would wait for months before any significant information or photographs could reach them from Kinshasa or abroad in the suitcases of travelers. It now takes just a few seconds.

The Internet also gave access to a number of press releases—from local, international or regional institutions, inquiry commissions—and other important data regarding the peace process in the region. Some national administrative bodies started developing communication strategies after years (and a very deeply rooted tradition) of withholding information.

Exchanging information from one country to another is especially sensitive in times of conflict between two states, or when a country is hosting on its terrain armed movements that are in rebellion against the government of a neighboring state. That situation prevailed in the African Great Lakes region for years, contributing to waves of xenophobia and nationalist upsurges in which the media were often involved. From 2004 onward, Congolese journalists have been able to access the Web sites of Rwandese newspapers or listen to Burundian radio stations online. Even if that is not sufficient to build a common understanding of the situation in the region, at least communication and dialogue are now technically easy.

It is thanks to these new digital opportunities that regional news supply projects, such as Syfia Grands Lacs Agency (http://syfia-grands-lacs.info), could be set up. This agency aims at providing to all media in DRC, Rwanda, and Burundi (and, by extension, other media in Central Africa or Northern countries), free stories written by journalists from those three countries. It is still rare to find an article by a Rwandese journalist published in a Congolese paper, but it happens at times. And it helps to undermine propaganda and stereotypes about the former "enemy."

Unfortunately, it has also become easy to "cut and paste" texts authored by others, in countries where there is almost no capacity to trace and prosecute abuses of that kind. It has become easier than ever to produce a newspaper just for the sake of promoting the views of a single individual or a political party, giving it the appearance of a real paper by "filling" the pages with other texts taken from the Internet. In a context where the journalists are usually lacking training, there is also a lack of capacity to criticize the credibility of the information circulating on the Internet.

A New Public Sphere

Just a few years ago, African journalists were physically very hard to locate and contact. Many newspapers and radio stations in Central Africa did not have a main telephone line operating[7] and it was a real challenge to reach a particular member of the team, especially the ones working in the field. Gradually, all journalists have obtained one, two, or even three cell phones (according to the number of operating networks in their area) and have also opened e-mail accounts. Journalists are now easily reachable by the audience or readers wanting to put forward their points of view, react to a piece of information published or broadcast, or willing to forward some confidential data. Thanks to the Internet, journalists not only can subscribe to mailing lists, information e-letters, and web-chats where important public matters are discussed, they can also receive messages directly from the public, making the traditional relationship between the media professionals and the audience much more interactive. Today in Kinshasa, the main newspapers (*Le Potentiel, La Référence Plus, Uhuru*...) indicate the cell phone number of the newsroom members in each edition.

It is also thanks to the cell phone that "talk shows" and "open-mike" programs could grow and give opportunities to more of the audience to make their voice heard. These programs have set the ground for the popularity of private radio stations in Burundi and are also very much appreciated in DRC and Rwanda. In the war situation, those programs provided opportunities to the population to make their voices heard by political leaders and armed groups. In Bukavu, for instance, Radio Maendeleo (Radio Development) gave airtime to war victims, including women who had been sexually abused, in the hope that the local authorities would take up their responsibilities and do something to prevent those abuses. Thanks to the cell phone, people from the villages around Bukavu could call the radio station to talk about the security situation in their area. In Burundi, private radio stations, such as Public African Radio (RPA) or Radio Isanganiro, dared to relay the accounts of civilians who witnessed abuses committed by the army (Palmans 2005, 71).

Though such programs are still "risky," as young radio presenters are not always able to control the expression of live callers, they symbolize a real freedom of expression for populations that were silenced for years, having been pushed into a position of passive receivers of top-down communication. The High Media Authority (HAM), the regulatory body of communications in DRC, has tried to adopt recommendations to ensure that professional principles are respected in "open-mike" programs and abuses limited.[8] In Rwanda, all private radio stations decided in 2005 to suspend for a few months all "open-mike" programs after a phone call by an extremist supporting the Tutsi genocide was broadcast live by one radio station (Frère 2006, 116–117). Thanks to the Internet and the cell phone, the capacity for rapid, anonymous, and hidden contacts with "informers" has grown, but the danger of manipulation has also increased.

Nonetheless, these developments are opening a new era for public participation in countries where radio pluralism did not exist until the beginning of the 1990s. A real new "public sphere" has been constituted, thanks to cell phone technology.

Producing News: Presenting Both Sides of the Stories

While multiplying the sources available for journalists, ICTs have also pushed media professionals in the region to better verify and cross-check information, and thereby provide more balanced, reliable, and rigorous news. When, after decades of state monopoly on media outlets, private media appeared in the euphoria of political liberalization at the beginning of the 1990s, many newspapers were "opinion" oriented, politically biased, and close to political parties or individuals. Impartial, neutral, and balanced reporting has very seldom been the main editorial commitments of these papers. In Burundi in1993, real "hate papers" (*Le Témoin*, *La Patriote*, *La Nation*) promoting ethnic hatred were circulated, as was the case in Rwanda in 1994 (Kaburahe 2004, 29). In DRC, each new newspaper seemed to be backed by a political party, or a politician, devoted to support its position unilaterally. When the war started, Congolese territory was occupied by several armed groups, and journalists could not travel from one area to another and were unable to do any fieldwork, and they could not therefore get information about what exactly was going on out of the government-controlled zone.

By facilitating the contacts between media and stakeholders (experts, witnesses, fighters), the cell phone gave more opportunities to the journalist to cross-check information and get access to contradictory views. In Burundi, thanks to the cell phone, the RPA could give voice to the rebel leaders as recently as 2001, making their claims more clear to the entire audience. In a context where it was extremely dangerous and unsafe to reach the rebels' area, cell phones enabled journalists to obtain their views about the clashes with the regular army. That pluralism within the programs, systematically presenting the side of the rebel as well as the position of the government, did not please the authorities. They tried for years to silence the private radio stations that would broadcast statements from the rebels, saying that those stations were "traitors" to their nation, for giving a voice to "killers" (Palmans 2005, 72). The openness of the private radio stations to all points of view, from the political and social field, was evidence of improved professional behavior. Technologies improved balanced reporting, aiding the growing popularity and credibility of private radio stations.

Distributing News: Getting on the Net

Since 2000, several major media in DRC, Rwanda, and Burundi have gone online. If those Web sites are probably not widely accessed locally, because of low connectivity, they have provided increased visibility worldwide. The coverage of African conflicts by the Northern press is always dominated by the perspective of Northern journalists. The Internet gives the opportunity to African journalists to present their points of view about happenings in their continent, region, or country.

For example, Radio Okapi in DRC can be heard on the web by the Congolese diaspora and by all those interested. The "peace" radio station, created in February 2001 jointly by Foundation Hirondelle (a Swiss NGO specializing in setting up peace media in conflict situations) and by the MONUC (the UN Mission in DRC), aims at covering the huge territory of Congo and at giving a voice to all the differing viewpoints. Thanks to technological progress (especially through satellite transmission), Radio Okapi reaches over 80 percent of the Congolese territory and has developed ten local production stations in the provinces. The radio was a key player in the process of "reunification" of the country, because it gave to all Congolese the information about what was going on in different parts of the country that had been (or still were) under rebel movements. Many observers acknowledge the fact that Radio Okapi played a large part in maintaining a sense of common belonging and common destiny among all Congolese, which was critical for the peace process. ICTs were of crucial importance in expanding the capacity of Radio Okapi, as the nationwide production centers exchanged sound files with the capital, Kinshasa, mainly through the Internet and cell phones.

Over time, the ICTs have also allowed journalists from the Great Lakes region to develop new types of media. In Burundi, where the written press faces huge economic difficulties (given the high cost of supplies in a landlocked country) and can hardly circulate publications regularly, online press agencies have multiplied. Alongside the government-controlled Burundian Press Agency (ABP), there are now four private agencies circulating news through the Internet: Net Press, Aginfo, Kirimba, and Zoomnet. These media are able to reach new audiences (especially the diaspora), and these audiences, who feel increasingly concerned about building peace, try to bring suggestions or contributions for conflict resolution based on their broad experiences.

Keeping the Memory of the Past: The Importance of Archives

African media have for a long time suffered from "amnesia" because of bad archiving conditions (for newspapers); while the poverty of the radio stations pushes them to endlessly reuse the same tapes. Digital technologies make it possible to keep track of the productions published or broadcast in an easy, cheap, and more sustainable format.

If the process is too recent to allow us to conclude that this "memory of the past" will be used regularly by journalists and change their daily way of working, it has already changed something in the way news producers understand their "social responsibility." In these three countries, communication regulatory bodies have been set up to ensure that, in a liberalized environment, private and public media would remain within the rule of law. Although they are very different in their ways of operating, regulatory authorities[9] have made the media professionals understand that they are held responsible for the papers they publish and programs they broadcast (Frère 2006, 117).

In DRC, Burundi, and Rwanda, media monitoring centers have been established, often within those regulatory bodies, in order to keep track of the content of publication and broadcast: the Monitoring Center of the Congolese Press in Kinshasa, Media Monitoring of the Organization of Media in Central Africa in Burundi, and High Council of the Press monitoring center in Kigali. Digital equipments allow those centers to easily store hundreds of hours of news programs. Copies have to be kept for reference in case a complaint comes to the regulator about a particular media content. Electronic archives are therefore not only important to daily news-production duty; they also help build a more responsible professional practice.

Protecting the Journalists and their Rights: The Power of e-Networking

In the Great Lakes region, during the war, the journalists have been regularly molested, arrested, and imprisoned, either by the government or by a different rebel movement. In Rwanda in 1994, more than half of the journalists were assassinated in the first days of the genocide (Chrétien et al. 1995, 351). During the war in DRC, forty-three journalists were arrested or put in jail in 1998, fifty-three in 1999, and forty-two in 2000 (AMI 2004, 52). The census elaborated by the organizations defending the rights of journalist has followed the escalation of violence during the armed conflict.

It is thanks to the ICTs that, in DRC, a network specialized in defending press freedom expanded and became a key player in the media landscape and one of the media organizations best known abroad. "Journalist in danger" (JED) was established in 1997, but, according to its President, Donat M'Baya, it could only fulfill its mission from 1999 onward after the Internet and cell phones came into play.

Before 1999, JED had to transmit its alerts and reports about journalists' rights abuses by fax, and the information could only be sent to a very limited number of targets. The fax transmission cost $2.5 per minute and sending a single press release lasted two minutes: that meant a US$5,000 expense to reach 1,000 receivers in DRC and abroad, which was unaffordable for a small NGO like JED (M'Baya 2004, 36). As early as 1999, JED was thus the first media organization in the country to develop its own Web site, hosted abroad, that really boosted JED's efficiency. At that very moment, for instance, Pierre Sosthène Kambidi from Radio Kilimandjaro in Tshikapa (West Kasaï), 1,000 km away from Kinshasa, was arrested and imprisoned. He wrote a letter to JED (at the time there were no cell phones yet in DRC), which took a whole month before reaching the JED office. As soon as JED staff received the letter, they sent out a press release about the case via the Internet. That same evening the information was broadcast on international radio stations (RFI and VOA). Three days later, thanks to that worldwide publicity, the journalist was set free (M'Baya 2004, 36).

Since 2000, mobile phones began to give JED much more efficiency for collecting data rapidly. JED has trained and provided its local correspondents in the provinces with cell phones and airtime vouchers. Two of the main worldwide organizations defending freedom of expression, the French Reporters sans Frontières and the Canada-based International Freedom of Expression Exchange, became partners with JED. They obtain and circulate JED alerts and reports in their own networks, giving a huge echo to JED's work on the field in DRC. Every single attack on press freedom committed on the DRC territory is immediately brought to the knowledge of thousands of electronic subscribers at very low cost.

It is also thanks to the Internet and the cell phone that JED had the capacity, in 2004, to launch a regional alert network that covers seven other countries. Unfortunately, even in 2008, the network remains weak. Connectivity is really low in countries such as Chad or the Central African Republic, and it can take up to a week before an inquiry from

JED about a particular case receives the appropriate answer from the correspondent.

What is the impact of such an organization on the daily security and working conditions of the media professionals? Though violence against journalists has continued even after the end of the war and all along the transition period in DRC, JED is now widely known and respected. President Kabila has in some cases responded to JED about their inquiries and suggestions on press freedom. Because of their popularity and the respect they get from local and foreign partners, JED started lobbying actively for "decriminalization" of press offences, in DRC and in the region, so that journalists would not be jailed for stories they have written or broadcast. The success story of JED is rooted in the power of networking, mainly through ICTs.

Training: Filling the Knowledge Divide

In conflict situations, the education sector is generally weakened or even ruined by the war. In DRC, the aspect of journalist training was already a difficult matter even before the war: the first journalism school created in the 1970s (the Communication and Information Sciences Institute/ Faculty, initially named ISTI, then IFASIC) was facing huge financial and operational problems. In Burundi and Rwanda, there was no journalism school at all: it was only in 1996 that the School of Communications and Journalism was created at the Rwanda National University in Butare, but with a perpetual lack of local teachers. ICTs have greatly strengthened the capacities of local trainers by giving them access to a wide range of knowledge materials.

Thanks to ICTs, media professionals can also get access to training sessions online (within TOFFRAACE, for example)[10] or to training material on CD-ROM. Search for Common Ground, a U.S. NGO that supports the participation of the media in "conflict transformation" or "conflict resolution" and "reconciliation," has developed several training sessions about important issues for journalists working in a conflict environment, which are now available online and can be used for "self-training" CD-R (SFCG 2006).[11]

Some NGOs have chosen to organize regional training, gathering journalists from Rwanda, Burundi, and DRC for the same training sessions, with the aim of fighting the distrust between them and promoting professional collaboration. The Panos Institute (Paris) initiated such a regional program in 2002, way before the end of the war, to bring together media professionals and civil society activists from the three countries in a process of mutual capacity building. Most of the participants in those training sessions keep in touch through e-mail. Following one of those training sessions, three radio stations from Burundi, DRC, and Rwanda launched a monthly joint program ("Regards croisés") in 2006 addressing the issues related to the consequences of the war, which is broadcast live simultaneously by the three media. Such a common program, which can gather an audience of millions of Rwandese, Burundian, and Congolese, could be set up only with the use of ICTs—especially cell phones

The Barriers: New Technologies, New Challenges?

The intrusion of ICTs in the journalistic field has not only generated positive consequences for the media in the Great Lakes but also some challenges, such as cases of manipulation, misinformation with poorly identified and unreliable sources, spreading of strongly biased news, and plagiarism from the Internet. The Internet has brought about access to such huge amounts of data that journalists do not always have the skill to sort what should be kept and what should be dismissed. There are two main challenges that still prevent journalists from the Great Lakes from making better use of ICTs to reinforce their professional skills: access to the net and "net-ethics."

The Challenge of Access and Cost

In the Central African countries emerging from wars, there is heavy damage and destruction of infrastructure, and the populations live in poverty and trauma. Access to the net is still limited in the newsrooms and many journalists are forced to check their e-mails in public cyber centers, paying with their own money. Therefore they do not regularly surf the web to get information for professional use-preferring to use their time on-line for personal matters. When there is an Internet connection in the newsroom, the facility is sometimes locked in the director's office and the journalists cannot easily use it. Connections are often slow and frequently interrupted by power cuts. Downloading heavy files (like long reports, training manuals, pictures, or sounds) can be a problem for under-equipped newsrooms. Generally, outside the capital city, expensive V-Sat technology is required to get access to the Internet, as the main phone lines cannot support a connection.

Related to these technical constraints are high access costs. In 2007, a twenty-hour monthly Internet access costed $85.95 in Burundi, $92.53 in DRC (2004),[12] and $79.74 in Rwanda (ITU 2008).[13] Cell phone access is also very costly. Press enterprises and radio stations usually operate in a situation of deficit budgets, and most of them do not provide air time for professional use to their journalists (Ouendji 2008). The only media that can do so are often the ones depending on funding by foreign donors or partners. For most private commercial media, all investment that does not immediately generate revenue is avoided. Moreover, no specific strategy toward implementation and ownership of ICTs in daily journalistic work is promoted. Setting up an "online" edition of a paper or a Web site for a radio station and getting archives in digital format incur expenses, including the cost of staff training, without any direct and immediate reward. Nevertheless, managers increasingly realize that the papers or radio stations that are "under-informed" face the risk of losing its audience, because, in the current context of pluralism, the consumer has now the choice to tune to other national or foreign media.

The Challenge of Net-ethics

ICTs are only conveyors of data and do not change the content, quality, credibility, or honesty of the news disseminated. Nor do they affect the professional rules that guide the journalistic methods and practices: access to more information, often from unknown sources, requires journalists taking on even more professional responsibility and making more balanced judgments.

In the Great Lakes region, during and after the wars that have torn down those countries, the journalistic fraternity has realized how media could play a part in conflicts and can produce tragic consequences by disseminating warmongering information. They have thus decided to adopt new codes of ethics and, in Burundi and DRC, to set up self-regulatory bodies that would monitor the media to make sure professional principles are respected. That is why the Observatory of Congolese Media was established in 2004, followed by the Observatory of the Burundian Press in 2005. The Rwanda Association of Journalists also set up a Commission for Ethics in 2006. These new organizations are inspired by examples of media councils in Europe, Northern America, as well as West and East Africa (the Observatory of Ethics and Press Freedom in Côte d'Ivoire and the Tanzania Media Council). Some of those self-regulatory mechanisms take into account the new challenges from the use of ICTs. For example, the Congolese Code of Conduct clearly addresses the problem of plagiarism. But these self-regulatory bodies do not compensate for the lack or the weaknesses of structures able to control and, if necessary, prosecute the misuses of Internet material (for instance, regarding copyright).

Ethics and codes of conduct are very often grounded in a common sense of "journalistic identity," a common understanding of what the journalist should be and the role he should play in a democratic society. In the Great Lakes region, the war has left a deep division among media professionals; many so-called journalists who showed that they did not care about idealistic journalistic principles chose to pursue different goals. For many media practitioners, the main aim has been to defend the political views of one of the war stakeholders or simply to make money, but not to provide balanced and neutral news useful to citizens. The result: "propaganda" or "survival" journalism, also fed by unverified Internet contents.

If the war situation had made a part of the fraternity to be conscious of the need to converge around a common regulated practice, including the "ethical" use of ICTs, it has also deeply divided the profession between this group and those using the media for purposes other than idealistic citizen-centered journalism. From abroad, Web sites animated by the Burundian and Rwandese diasporas are still circulating "ethnic hatred," widening the gap between communities, and denying the occurrence of the Tutsi genocide. In Bujumbura, in July 2003, the director of a local online press agency, Net Press, was arrested because his agency provided a link to the Agora Web site, hosted in Denmark, which was circulating violent propaganda against the transition president, Ndayizeye, and his community. In 2002, the regulatory authority had prevented national Web sites from reproducing any content published elsewhere threatening peace and security or praising violence. During the 2006 election in DRC, hate speech against the international community, precisely the MONUC and Belgian people, calling for the boycott of the polls, was spread through the Internet from abroad and sometimes relayed by the local press in Kinshasa. Thanks to ICTs, the venom of hate speech, threatening peace and stability efforts inside the region, can easily be distilled from abroad, and local journalists must stand by their professional code of conduct to bar its expansion: if the region is in flames again, local journalists will be the ones enduring the consequences.

Conclusion

Without the ICTs, it is doubtful that the Great Lakes media could have played a positive part in the conflict and in peace-building. More than any training efforts to promote "peace" or "conflict-sensitive" journalism among local media, it is the ICTs that have changed the professional practice of journalism. The new opportunities they brought to journalists in that landlocked area certainly contributed to a better participation of the media in peace efforts. However, ICTs are just a technical opportunity. To be used in the best manner (be it peace-building or any other field), "they require skill levels and commitment that often are simply not there," noted the ITU 2006 Report. They can be instrumental to the journalists who want to balance information, obtain complete data on events to be shared with the citizens, or lend an ear to the feelings and understandings of his or her audience. But the ICTs can also be useful to the warlords and their mouthpieces who want to manipulate the minds of people or spread terror in the community. The war in the Great Lakes area has produced examples of both, although one can assume that the positive consequences of journalists' use of ICTs outnumber the negative ones. If journalists become more aware and professional by using ICTs, they may gain a maturity that could place them in a position to promote democracy in ways that go beyond the political elite. If they attain a level of sophistication that outstrips even that of the politicians in their own countries, they could help to build the road to peace.

Notes

1. The data from 2004 are taken from the ITU report: ITU, *World Communication/ICT Development Report 2006*, Geneva, March 2006. The 2007 data from ITU (*African Telecommunication / ICT Indicators 2008: At a Crossroads*, May 2008) show that the growth is continuing, with 2.94 cell phone users for one hundred inhabitants in Burundi, 10.52 in DRC, and 6.98 in Rwanda. Internet users reach 0.77/100 in Burundi, 0.37/100 in DRC, and 1.08/100 in Rwanda.
2. To compare with Northern countries, the ratio of cell phone subscribers was, at the same period, 73.72 for 100 inhabitants in France, 102.16 in the UK, and 62.11 in the United States.
3. The same report mentions 41.37 users for 100 inhabitants in France, 62.88 in the UK, and 63 in the United States.
4. That is why after the 1994 genocide in Rwanda, BBC and VOA launched programs in Kinyarwanda and Kirundi (an hour a day).
5. Colette Braeckman, interview, December 16, 2006.
6. See the Report of the Bujumbura workshop, "Media and citizenship in Central Africa" (December 9–11, 2006), organized by the Panos Paris Institute.
7. The number of main telephone line users in those countries is now lower than the number of mobile phone subscribers. In 2007, it reached only 0.45 line users for 100 inhabitants in Burundi, 0.02 in DRC, and 0.18 lin Rwanda (which represented an evolution of respectively +9.2 percent, –3.7 percent and –12.6 percent for the period 2002–2007) (ITU, 2008). In developed countries, the ratio is 56.04 main line users for 100 inhabitants in France, 56.35 in the UK, and 60.60 in the United States (2006 ITU Report).
8. The Decision n°HAM/AP/002/2004 on open-mike radio and television programs states clearly that these programs must respect journalistic ethics and that the presenters must be experienced media professionals.
9. HAM (High Authority for the Media) in DRC, CNC (Council for National Communication) in Burundi, and HCP (High Council of the Press) in Rwanda.
10. TOFRRAACE (meaning "All training opportunities for community, educative and association radio stations in Africa") is a

program that was initiated by three community radio associations based in Burkina Faso, Benin, and Kenya. After setting up a virtual resource centre for community radio (TRRAACE: http://www.mediafrica.net), they launched online training sessions for African journalists and radio managers in 2006 (http://tofrraace.mediafrica.net).

11. Search for Common Ground published several guides that are easy to download from their Web site (www.sfcg.org): *Radio Talkshows for Peace Building* (by Ross Howard), *Youth Radio for Peace Building-a guide* (by Michael Shipler), *Soap Opera for Conflict Prevention* (by Francis Rolt).
12. No data are available for DRC after 2004.
13. During the same period, in 2006, twenty hours of internet connection in developed countries (which have a level of revenue much higher than the Great Lakes countries) cost $ 31.29 in France, $29.02 in the UK, and $14.95 in the United States.

References

AMI (African Media Institute). 2004. *Les entraves à la liberté de la presse en RDC*. Kinshasa.

Baumann, M. and Siebert, H. 2000. "Les journalistes comme médiateurs." In Reychler L. and Pafenholz, T. *Construire la paix sur le terrain: mode d'emploi*. Brussels: GRIP-Complexe.

Berger, G. 2005. *Doing Digital Journalism. How Southern African Newsgatherers are Using ICT.* Highway Africa, Rhodes University: Grahamstown.

Chrétien, J. P. et al. 2005. *Rwanda: Les Médias du génocide*. Paris: Karthala, 1995.

Frère, M. S. 2006. "Après les médias de la haine. La régulation au RDC, au Rwanda et au Burundi." In *L'Afrique des Grands Lacs. Dix ans de transitions conflictuelles, Annuaire 2005–2006*, ed. F. Reyntjens and S. Marysse, Paris: L'Harmattan.

Frère, M. S. 2007. *The Media and Conflicts in Central Africa*. Boulder: Lynne Rienner Publishers.

Howard, R. 2003. *Conflict Sensitive Journalism*. International Media Support and IMPACS, Vancouver-Copenhagen.

Institut Panos Paris. 1998. *Internet à l'usage des journalistes africains*. Paris: Karthala.

International Telecommunication Union (ITU). 2006. *World Communication/ICT Development Report 2006*. ITU, Geneva.

International Telecommunication Union (ITU). 2008. *African Telecommunication/ICT Indicators 2008: At a Crossroads*. ITU, Geneva.

Kaburahe, A. 2004. *La mémoire blessée*. Bruxelles: La Longue Vue.

Lynch, J. and McGoldrick, A. 2001. *Peace Journalism. How to Do It*. Transcend, Sydney/London.

M'Baya, D. 2004. "Les TIC et la promotion de la liberté de la presse." In *Final Report of the First Festival of the Media in Central Africa*. Bujumbura, 3–5 mai 2004, OMAC (Organization of the Media in Central Africa).

Ouendji Ngatta, N. 2008. Les journalistes à l'heure du téléphone portable: Usages et enjeux dans les médias camerounais, Ph.D. Thesis. University of Bordeaux III.

Palmans, E. 2005. "Les médias face au traumatisme électoral au Burundi." In *Politique africaine* n 97. mars 2005, Paris, Karthala.

Thompson, A. 2007. *The Media and the Rwanda Genocide*. London: Pluto Press.

Zacharie, A. and Janne d'Othée, F. 2004. *L'Afrique Centrale dix ans après le génocide*. Bruxelles: Editions Labor.

CHAPTER 14

Conflict Coverage in a Digital Age: Challenges for African Media

Rune Ottosen and Okoth Fred Mudhai

Introduction

In an increasingly networked and globalized—or globalizing—society, the recent emergence of new information and communication technologies (ICTs) in a fast-pluralized media ecology presents challenges for most of Africa's hitherto state-dominated "big man" politics. We argue in this chapter that in situations of political-ethnic conflicts, new digital tools produce opportunities for propaganda, but at the same time they also offer new possibilities for counterpropaganda. Alternative information from blogs, e-mail lists, Web sites, especially through NGOs with Web sites and e-mail networks, have—in addition to cell phone text messaging—complicated recent political contests linked to socioeconomic tensions. This chapter not only analyzes the links between violence and electoral politics but also examines the contribution of new ICTs in exacerbating or ameliorating inter-"ethnic" and interparty violence—especially around national polls in selected African countries.

Media Coverage in Modern Africa

For about three decades in postcolonial Africa, media—especially broadcast—"coverage" of news was characterized by the activities and words of the "big man." From the 1990s, public access to information increasingly moved away from the centralized newsroom "docket" or diary, thanks to new ICTs. This marked a change in communication culture. As Castells (2002, 139) puts it, "the internet is not simply a technology: it is a communication medium (as the pubs were)...the internet became the indispensable component of the kind of social movements emerging in the network society." In Africa, as elsewhere, this networked society offers new challenges as well as new opportunities not only through transnational radio and satellite TV but also through cell phones as well as blogs and other forms of independent distribution of content on the Internet. Indeed, when it comes to conflict coverage, the exponential increase in Internet traffic "underscores its expanding importance and the need for it to be reevaluated as a topic for news coverage...that journalists, among others, rely on so heavily" (Seib 2004, 103). For this reason, journalists and other communicators need to be aware of the nature of such new ICTs, that of a double-edged sword, and adjust their news coverage to the sensitivities of the moment and context.

Such understandings need to take into account local-national, regional, and global developments. The network society is part of the globalization process, underlining the McLuhannian "global village"—more recently captured by Robertson's notion of the world becoming "one single place" (Robertson 1992). One of the many challenges for African media is the impact new ICTs are having on national and regional conflicts such as those in Sudan and Congo (Orgeret 2006). This is more so given that the Internet can be a "weapon," in fact "it must be looked upon as a tool of war as well as an information tool" (Seib 2004, 103). Recent eruptions of brutal violence not only in the Zimbabwean 2008 elections but also in the erstwhile more stable Kenya (over the December 2007 presidential elections) and in South Africa (over immigration) indicate a trend of conflicts-other-than-war approach being increasingly deployed to solve political and socioeconomic problems. The media in general and ICTs in particular play a role in these situations.

Theoretical Approaches on Conflict Coverage

Peace journalism has been suggested by authors such as Galtung (2002), as well as Lynch and McGoldrick (2005), as an alternative strategy for critical journalists covering wars and conflicts. Galtung's model builds on the dichotomy and the contrast between what he calls "war journalism" and "peace journalism" approach. The model includes four main points that contrast the two approaches: war journalism is violence-oriented, propaganda-oriented, elite-oriented, and victory-oriented. This approach is often linked to a zero-sum game where the winner (as in sports journalism) takes all—a prototype of what one might call traditional mainstream war coverage, without the journalists reflecting that media themselves are playing a role in the conflict, often escalating it by reproducing propaganda developed as a part of media strategies and PR campaigns by the parties involved.

BBC reporter David Loyn is the most known journalist openly opposed to the peace journalism approach. In a special issue of *Conflict & Communication Online*, in which the opponents and defenders of peace journalism debate the issue, Loyn prefers to use terms such as "truthfulness" and "objectivity" as journalistic guidelines, even though he acknowledges their limitations. "On this analysis, if we accept that objectivity is at least a worthy aspiration, even though not a tool to achieve the 'whole truth,' then peace journalism fails a key test by imposing other expectations onto journalists" (Loyn 2007, 5). Loyn is disturbed by Galtung's original model, whose categories he views as too dualistic. He claims that everyday news journalism seldom fits into "war journalism" or "peace journalism" categories. Loyn underestimates the power of propaganda. A reporter stepping into a conflict arena can easily be part of the problem and escalate violence if he or she does not acknowledge that journalists play a role regardless of their subjective ambition to be neutral. This is because the parties involved in a conflict have already "infected" the conflict with propaganda, hate speech, enemy images, and prejudices before the journalists arrive (Ottosen 1995; Nohrstedt and Ottosen 2001). A peace journalist knows this out of experience. As a professional journalist he or she does not report uncritically when a source claims that "the other" threw the first stone. He or she has already studied the historical background and knows that sources from the other side have similar claims of the stone-throwing. As Lynch and McGoldrick put it, a peace journalist has no illusion that he or she is above the conflict and acknowledges that the journalists are in the "boat together with the conflict" as soon as they physically step into the war zone. Only when you accept *that* fact, can you deal with the problem and cope with it in a professional manner. And who knows, journalists might even make a difference and promote peace through their reporting by highlighting some issues that the victory-oriented parties want to hide.

Thus, the peace journalism approach has a moral and ethical point of departure, acknowledging the fact that media themselves play a role in the propaganda war (Lynch 2007). It presents a conscious choice: to identify other options for the readers/viewers by offering a solution-oriented, people-oriented, and truth-oriented approach, and this in turn implies a focus on possible suggestions for peace that the parties to the conflict might have an interest in hiding. Peace journalism is people-oriented in the sense that it focuses on the victims (often civilian casualties) and thus gives a voice to the voiceless. It is also truth-oriented, in the sense that it reveals untruth on all sides and focuses on propaganda as a means of continuing the war (Galtung 2002, 261–270). However, the concept of peace journalism will hardly be interesting for journalists in Africa unless issues like the digital divide are included in the analytical perspective (Ottosen 2007).

The Digital Divide as Expressed in the "Electronic Colonialism Theory"

A critical approach to war and conflict coverage should focus more on technological and economic issues (Ekecrantz 2007). A critical look at the development in the media industry should include a global economic perspective in which neoliberalism sets the cultural agenda. Miller (2007) introduces the notion of "cultural citizenship" as a tool for analyzing the role of media as an arena for discourse on development and peace issues. "For cultural studies, cultural citizenship concerns the maintenance, development, and exchange of cultural lineage—a celebration of difference, which is also a critique of the status quo" (Miller 2007, 23). However, only a small minority is able to make use of the new digitized publishing reality.

The Internet, a network that began only in the 1990s, has spread globally, but with uneven speed. There are huge differences in its spread *within* the developing world. While a 2003 Goldman Sachs report indicated that over the next few decades Brazil, Russia, India, and China would increase their share of the world economy (Thussu 2007, citing UNCTAD 2005, 15 and 153)—partly driven by ICTs,—Africa remains far behind the developed world due to its peripheral role in the global media economy, including with regard to ICT access (Gray and Esperanza 2004). If we look at the global share of the value of exports of selected media products such as newspapers and periodicals, Europe and North America have 94 percent of the total export value, Asia 3 percent, and Africa 0.1 percent (Thussu 2007, citing UNCTAD 2005, 15).

Technological development runs parallel with signs of uneven regional development. In regions of the developing world, such as the Middle East and Northern Africa (MENA), we see an increase in Internet use and new media reaching larger local audiences. Thus we can see tendencies to a digital divide also within the African continent.

In Africa there are significant regional and national differences in the level of technical development, media use, and development of professional journalism. Rønning (2005) points out that, on one hand, globalization of the media helps in promoting democratic ideas and liberal thinking in many African countries. Obviously, there are positive trends in many countries regarding the freedom of expression and technological development in the media industry (Teer-Tomaselli et al. 2007). On the other hand, the situation in countries such as Zimbabwe is grave for independent journalism. Through constant surveillance on the Internet, the banning of Internet publications, the use of Internet filters, and even—in some cases—imprisonment of publishers, Mugabe's regime has challenged all independent voices. Thus the Internet is not in itself a guarantee of more democracy. In Zimbabwe, repressive media laws are implemented to stifle media and publishers (Rønning 2005, 165). South Africa is also facing threats to press freedom, legal proposals that might result in censorship (*Svenska Dagbladet* August 1 and 5, 2006). Even though the Internet makes it easier to promote democratic ideas and counterpropaganda against authorities, the information gap within Africa itself and within each African country is a challenge. Rønning puts it thus:

> Even if information technologies are being developed in Africa to a substantial degree, they are still very much a minority phenomenon and while many African newspapers now have electronic editions (with all 54 African countries now having internet access), this does not mean that they reach more than a limited part of the population. The information and media gap between the information-strong and information-weak parts of the world is not going to disappear despite optimistic preannouncements of the internet creating...democratic communication in the near future (Rønning 2005, 165).

Tendencies toward a digital divide within Africa can be partly seen against the background of U.S. public diplomatic attempts aimed at supporting democratic ideas and modern mass media (Figenschou 2006). Many critics claim that this process must be seen as part of the struggle for U.S. hegemony over the region, a civil parallel to the military expansion. In Somalia, the only Somali Internet service provider connecting the Somali diaspora was forced to shut down after pressure from the U.S. government because of suspected links with terrorists (Georgiou and Silverstone 2007, 37). Leaving that discussion aside, there is no doubt that the expansion of modern communication and mass media in MENA has been substantial.

> Nowhere has international media taken off so rapidly in recent years as the MENA, home to 480 million Arabs, Jews, Africans and Persians in the broadly defined 26-country region that mainly follows the Mediterranean from the Straits of Gibraltar to Turkey, and from the Red Sea to the Indian Ocean down to the Comoran islands. In percentage terms the region is one of the world's fastest growing places for adopting the mechanisms of international communication: satellite dishes, mobile telephony, computers, televisions, and internet service providers (Berenger 2006, 193).

However, Berenger points out that many of these societies have not modernized their political systems. They still lack democratic reforms, in common with many countries in sub-Saharan Africa, which have not developed technologically. Hence, there is no direct correlation between the development of a modern media and a country being politically reformed (Berenger 2006, 193–195).[1]

Democratic Gains and Losses: New Technology, Empathy and Professionalism

Obviously, the Internet offers a new alternative to both established media and oppositional voices. The reports of NGOs such as FAIR[2] and War and Peace can distribute alternative information that challenges the information monopoly highly influenced by propaganda and psychological operations (Taylor 2007).

> Through these and other independent sources, the internet dilutes the potency of governments' "official news," which is most effective when it faces minimal competition. The voices of the web are obstacles to the flow of propaganda because they provide so many information options... Web-based information sometimes provides an alternative to conventional wisdom, and it may enhance citizens' ability to do more—to participate in political action (Seib 2004, 95).

While "the internet's future as a battleground may be uncertain... among its most significant functions is to serve as one of the tools being used throughout the world to give voice to those who long have not been heard" (Seib 2004, 103). Individual bloggers can sometimes make a difference, facilitating direct communication and sharing a common concern about the consequences of the war and political conflicts to the civilian population in each country (Fengler 2008). Of course, individual voices can never reach the same audience as established media. In any case, even mainstream media can sometimes use digital technology to achieve remarkable results. This aspect of peace journalism is obviously underestimated in the debate on the subject (Ottosen 2007). Let us now highlight and examine instances in which, from recent literature and our observations, ICTs have been deployed not only by journalists but also by ordinary local and diaspora citizens during conflict situations, whether for good or evil.

Citing the argument by Morris and Ogan[3] that resorting to the use of cyberspace to obtain information and data reaches its peak during events with highly emotive collective values, Guazzini (2001 and 2004) surveyed fifteen Web sites that "covered" the 1998–2000 military conflict between Eritrea and Ethiopia. The author's observations include the following: "formalized and sophisticated manipulation" of information on the Web sites of both protagonist governments; identity politics in the "cybernetic war propaganda" targeting national public opinion; a number of unofficial Web sites with some semblance of neutrality alongside those that are openly partisan; the peaks in Web site access coincided with the most serious outbreak of battles; many of the users were based abroad; cyberspace was also used to encourage peace initiatives. Often, developments in the real world of this conflict were subjects of intense web discourses.

> From 1999, the internal disarray grew even in cyberspace and revealed itself openly in May 2001, when 15 members—including ministers and army generals—published online a letter opposing the centralised management of power as it exists under Issayas Afeworki . . . Cyberspace continues to act as a megaphone for the regulations issued by the two government blocks, but it also serves to voice internal dissent and to air debates of an internal political nature, not only and not always serving as a space for antagonistic discussion (Guazzini 2004).[4]

Although Guazzini's study was limited to Web sites only and was carried out in the early days of the Internet, it gives some indication of trends in the use of the web during armed conflict in Africa. It shows that new media can bring into the mix more diverse discourses that are more difficult to monitor and control, unlike the more institutional forms of media and communication. Whereas the author reports that "journalists from both countries have agreed on ways to propagate news which might help to seal the fracture caused by the conflict, acting in such a way as to reestablish an atmosphere of reciprocal understanding and co-operation" (Guazzini 2004), discord in the Web site content shows that such an arrangement with new media users was not easy or possible.

A similar situation could be discerned in Kenya when mainstream media managers and journalists at some point agreed to preach peace[5] following the country's most vicious postelection violence in late 2007 and early 2008, yet messages of anger, acrimony, and hate continued to be distributed via the web, e-mail, and cell phones. The Kenyan electoral violence provided the launching ground for the Ushahidi (Swahili for "witness") platform[6] for crowdsourcing crisis information—later used xenophobic violence in South Africa and deployed in the Democratic Republic of Congo in November 2008. Following the violence that within just a few weeks resulted in the death of thousands of people and displacement of millions, researchers for the BBC World Service Trust carried out—among others—"an examination of claims that blogs and SMS text messages were used to inflame tension and incite ethnic hatred"

(Abdi and Dean 2008, 3). The new (ICT) media, previously hailed for their role in facilitating democracy and economic development in Kenya, became harbingers of doom—going by reports of the country's two main human rights NGOs (Abdi and Dean 2008, 10). The site manager of a popular Kenya-focused blog had to shut it down at one point "because we haven't managed to control the heinous messages" (Abdi and Dean 2008, 11). On their part, cell phone provider Safaricom resisted official pressure to shut down the SMS system along the lines of the Ethiopian government's move to bar text messages in their 2005 election (Abdi and Dean 2008, 11). Before long, the Kenyan Parliament voted to amend the Kenya Communications Act to introduce mandatory requirement for cell phone subscribers to register with their service providers because "untoward people used mobile phones to sow seeds of discord" (Opondo and Orlale 2008). Indeed, the state charged in court one user identified as the originator of a chain hate message (Thuku 2008).[7]

To blame ICTs solely for the intensification of the Kenyan political unrest would be to subscribe to the "powerful" media effects theories that are difficult to prove. Many other variables, including the pronouncements of politicians at public rallies, played some role. In any case, many of those who resorted to the use of blogs and SMS to obtain details of various happenings or to pass on information did so to counter official propaganda—especially when the government banned the mainly local broadcast media from live coverage of election news. Although the discourses about the Kenyan crisis may not have strictly taken the form of what Hamelink (2008) terms "disarming conversation," they included choices for resolution. "Conflict is not inherently negative. All processes of change involve conflict" (Hamelink 2008, 296).

Whereas images of Kenyan violence in the international public sphere may have contributed to a resolution of the electoral conflict through a power-sharing arrangement, these CNN-effect images did not have as much impact on the 2008 Zimbabwean electoral crisis. All the same, given the role of "new media technologies as soft power tools . . . in what Joseph Nye terms a 'smart' approach to conflict" (Conway 2008, 236), Zimbabweans did not give up the new media battle against Mugabe's regime. One example of " 'smart mobbing'—galvanizing mass political action electronically . . . a carefully planned deluge" (Seib 2004, 96) was the Save Zimbabwe from Mugabe e-petition calling on South African President Mbeki and other Southern African leaders to hold an emergency summit to resolve Zimbabwe's electoral conflict.[8] Its target of 100,000 was reached within two days. Attempts by Mugabe's regime to stifle free expression in the new media spheres did show his irritation, but little evidence existed to show that the threat of this "soft power" was enough to result in any significant political change. However, pressure from these online and off-line quarters resulted in a sham rerun of elections and also in Kenya-style negotiations, leading to a Mugabe-Tsvangirai power-sharing negotiations from September 2008.

Another country in which attempts were made to use violence to solve sociopolitical problems is South Africa, where local youths attacked immigrant workers and residents from other African countries in 2008. In comparing the Kenyan and South African violence, Kimemia (2008) partly pointed accusing fingers at the liberal communication environment enhanced by new media.

> Just as in Kenya, South Africa is not entirely short of political operatives who... mobilise a section of the population around emotive issues... In South Africa, just as in Kenya, greater democracy appears to have emboldened every other miscreant into "democratically" expressing their dissatisfaction even if such expressions invariably involve breaking the law. With greater democracy in these two countries, there has been an inadvertent liberalisation of violence... (Kimemia 2008).

A Kenyan economist, who was programme manager for a South African NGO, Kimemia (2008) pointed out that the attacks took place a year before South Africa's general election—thus linking violence to electoral politics.

With regard to Somalia, Brinkerhoff (2006, 25, 26), while agreeing that "the internet can serve as a tool for nefarious purposes," takes a more positive approach—emphasizing how (Internet-organized) digital diasporas can "promote liberal values, channel frustration into verbal debates thus diffusing tension, and create communities that counter the marginalization conducive to violence." The use of only a single network group as a case study is limiting, and the author also acknowledges that "the potential impact of digital Diasporas in preventing or deflecting conflict... cannot be measured" (Brinkerhoff 2006, 27). However, the findings indicate, among others, that "Somalinet members seem to have a tendency to discourage destructive means of violence to bridge their differences rather than inflame them" (Brinkerhoff 2006, 43). The conclusion is that "both the internet and diasporas cannot be assumed to be universally threatening to home country stability" (Brinkerhoff 2006, 46). The author ends with a cautionary note with regard to the influence of such organizations. While the research leaves the "impact" question wide open, its tone is in tune with proposals by Hamelink (2008) on a more positive approach to discourses in the media and communication realms in times of conflict.

In sum, the examples highlighted above indicate that most conflicts in Africa are not only about resources but also about protagonists battling with each other over their points of view on the dispute. Victory in political contests tinged with violence is about "whose story wins" in "noopolitik," which includes "netpolitic," according to some authors (cited in Conway 2008). Media and communication play significant roles in peddling stories about victors and losers in a conflict. This may lead to escalation or resolution of the conflict. The arena provided by the new media often complicate the discourse environment.

Future Challenges for Research on Conflict Coverage in African Media

One of the huge challenges for African media is to face the relationship between conflict resolution and peace keeping, given that the UN model has failed in the "war on terror." The Norwegian political scientist Øivind Østerud is among those who have documented that the most violent and conflict-ridden areas are those places where the United States has forced through elections without there being any established democratic institutions or free media (Nohrstedt and Ottosen 2007). According to Østerud there are two lessons from this. First is that organizing elections in developing countries does not automatically lead to democracy and stability. Even if you remove a dictator through military intervention, corruption and undemocratic structures can occur in new forms. Elections thus could lead to a democratic fiction. Second, semi-democracies tend to be more vulnerable to violence than authoritarian systems and real democracies. Semi-democracy causes resistance and has no power to resist violence. There is no evidence that, in the long run, forced elections and semi-democracies will serve democracy, since anarchy and violence replace the relative stability that an authoritarian system produces. Even though this model has failed in Iraq, Afghanistan, and the Middle East, it is still the chosen model for the UN in Africa.

Even though this model failed in the former colonies, NATO, the UN, and the EU insisted on following this path in Bosnia and Kosovo, and Africa. This mild form of imperialism has produced no lasting internalized democracy. Still, it seems that this is the chosen policy for the UN in both Sudan and Congo. In Congo, we saw 20,000 soldiers from the EU and the UN, together with thousands of people from NGOs equipped with billion-dollar budgets, making an attempt to administer the country. The situation seemed calm on the surface after the elections in 2006, but underneath the surface there was still instability and violence (Østerud 2007). What role can media play in a situation like this? It is difficult to see the media playing an independent role, given the huge economic interests involved. Congo is resource-rich, providing a huge potential for earning profits, and the humanitarian explanation notwithstanding, economic interests remain a major factor. There is obviously great-power rivalry for hegemony in Africa—China, the United States, and the EU. It is difficult to see a UN policy in Africa independent of these interests. It is also difficult to visualize a media developing independently of these interests.

In many African countries there is a well-justified suspicion of foreign intervention in the name of democracy, since the parallel to the rhetoric of their colonial past is obvious (Omaar and de Waal 1993). At the same time, the rhetoric of Mugabe that justifies attacks on media and democracy in the name of the anticolonial struggle has proved to be disastrous (Rønning 2005). In his research on the link between cities, conflict, politics, and the media, Hamelink (2008, 291) suggests that "communication in the sense of 'disarming conversation' can play a crucial role in preventing low-intensity disputes escalating towards violent and deeply damaging conflicts." However, his concept is not easily achievable in a cacophonous new media environment.

Concluding Remarks

The role of new ICT media during conflict in Africa has yet to develop to the levels of what Seib (2004, 98, 110) terms "cyberconflict" or "conflict on the internet [which] involves weaponry...viruses, worms, and direct attacks." All the same, African journalists and media must be aware of these potentials and find their own way of covering future conflicts in a responsible and constructive manner. Unless the digital divide is dealt with, peace journalism and other alternative approaches to journalism will have little effect on everyday media use in Africa. On the ideological level, the concept of peace journalism could be an inspiration to African journalists, but without a stronger focus on the technological aspect of media development, the model will remain irrelevant for African journalism.

The Internet offers new possibilities of finding new paths through which to communicate outside the sphere of mainstream media. There are sharp regional differences in Africa when it comes to development of media technology (Teer-Tomaselli et al. 2007). In many African countries we have to acknowledge that ordinary people even now lack access to new digital media, and the flow of information is still influenced by the digital divide within Africa (Berenger 2006).

New technology, such as a cell phone with a camera, can be a powerful tool in spreading news about conflicts and human disasters. Examples of enterprising use of not only the cell phone but also the e-mail and the web in recent conflict situations show that much more could be expected of the more advanced cell phone handsets as they get cheaper.

The world has become a smaller place, in the sense that communication through digital technology, and the Internet in particular, helps in sharing information on a global scale (Clayes 2006; Castells 2002)—although the technology is still dominated by economic power centers (Thussu 2006). Truth-seeking journalists and alternative media can challenge these power centers through the counterflow of information, including peace journalism and other alternative strategies (Hackett 2006). But in order to use the potential of new digital media, the first priority should be on the implementation of new media and digital technology throughout Africa. The process is in place already, but it remains to be seen how far new broadband infrastructure will make ICTs more accessible to the general public and to what extent such new media would be deployed to further conflict resolution rather than to foment hostility.

Notes

1. Africa has also her hands full with other problems, including political instability, rebel wars, the threat of HIV/AIDS, other health issues, and poverty, which continue to take precedence over African media coverage of the "war on terror." As an example, Washington mentions the presentations by the Truth Commission in Sierra Leone that took place at the same time as the media in the North were focusing all their attention on Iraq. Some 50,000 people died in that civil war, but the Commission itself was an important historical event almost unnoticed by U.S. and European media (Berenger 2006, 1993–1995).
2. See: www.fairreporters.org
3. Morris, M. and C. Ogan. 1996. "The Internet as Mass Media." *The Journal of Communication* 46(1): 39–50.
4. Citing newspaper and radio reports in the local and German media.
5. One day, the media simultaneously printed and broadcast peace messages. At other points earlier in the conflict, speculation was rife that private, mainly vernacular, radio disseminated hate messages targeting certain ethnic groups because of their political leaning. See also Abdi and Dean (2008).
6. For details, see: www.ushahidi.com (accessed December 27, 2008). The platform, linked to Google Maps, enabled ordinary citizens to submit crisis information through text messaging via cell phones, e-mail, and the Web.
7. See also "Kenya charges man for SMS hate speech." http://inanafricanminute.blogspot.com/2008/05/kenya-charges-man-for-sms-hate-speech.html "Kenya: SMS the new guns of war." http://en.afrik.com/article12629.html (both accessed September 3, 2008).
8. See http://www.avaaz.org/en/save_zimbabwe/98.php?CLICK_TF_TRACK (accessed July 7, 2008).

References

Abdi, J. and J. Deane. 2008. *The Kenyan 2007 Elections and their Aftermath: The Role of Media and Communication*. London: BBC World Service

Trust. Policy Briefing No. 1. http://downloads.bbc.co.uk/worldservice/trust/pdf/kenya_policy_briefing_08.pdf (accessed June 25, 2008).

Berenger, R. 2006. "Media in the Middle East and North Africa." In *Global Communication: Theories, Stakeholders and Trends*, ed. T. L. McPhail, 192–225. 2nd edn. Boston: Blackwell.

Brinkerhoff, J. M. 2006. "Digital Diasporas and Conflict Prevention: The Case of Somalinet.com." *Review of International Studies* 32: 25–47.

Castells, M. 2002. *The Internet Galaxy: Reflections on the Internet, Business and Society*. Oxford and New York: Oxford University Press.

Clayes, L. 2006. Internet in Everyday Life: Boundary Conditions for Equal Opportunities in the Network Society. Paper Presented to the Digital Divide Group at the IAMCR 25th Conference, July 26–30, 2006, Cairo, Egypt.

Conway, M. 2008. "Terrorism and the Making of the 'New Middle East': New Media Strategies of Hizbollah and al Qaeda." In *New Media and the New Middle East*, ed. Philip Seib, 235–258. New York and Basingstoke: Palgrave Macmillan.

Ekecrantz, J. 2007. "Media and Communication Studies Going Global." In *Media Structures and Practices As Time Goes By*, ed. U. Carlsson and K. Helland, 169–181. Gothenburg: Nordicom.

Eriksen, T. 2000. Globalization, Winners and Losers. Unpublished paper. Oslo: SEFIA-Høgskolen i Oslo.

Fengler, S. 2008. "Media, Journalism and the Power of Blogging Citizens." In *Media Accountability Today and Tomorrow*, ed. T. von Krogh, 61–68. Gotenburgh: Nordicom.

Figenschou, T. U. 2006. "Courting, Criticism, Censorship and Bombing: The Bush Administration's Troubled Relations with Al-Jazeera Channel from September 11 to the War in Iraq." *Nordicom Review* (1): 81–96.

Galtung, J. 2002. "Peace Journalism—A Challenge." In *Journalism and the New World Order, Vol.2: Studying the War and the Media*, ed. W. Kempf and H. Loustarinen, 259–272. Gothenburg: Nordicom.

Georgiou, M. and R. Silverstone. 2007. "Diasporas and Contra-flows Beyond Nation-centrism." In *Media on the Move: Global Flow and Contra-flow*, ed. D. K. Thussu. Milton Park: Routledge.

Gray, V. and E. Magpantay. 2004. "Overview of Global Trends." In *From the Digital Divide to Digital Opportunities*, ed. G. Sciades. UNESCO.

Guazzini, F. 2001. "Riflessioni Sulle Identità di Guerra nel Cyberspazio: Il Caso Eritreo-etiopico." *Africa* LVI (4): 532–572.

———. 2004. The Construction of Political Nationalism on the Web: Borders, Social Spaces and Identities in the Ethio-Eritrean Case Study. Paper Presented at the International Conference on States, Borders, and Nations: Negotiating Citizenship in Africa, May 19–20, Centre of African Studies, University of Edinburgh, Scotland, UK.

Hackett, R. A. and W. R. Carroll. 2006. *Remaking Media: The Struggle to Democratize Public Communication*. New York and London: Routledge.

Hamelink, C. J. 2008. "Urban Conflict and Communication." *International Communication Gazette* 70(3–4): 291–301.

Kimemia, N. 2008. Where Life is Nasty, Brutish and Short. *Sunday Nation*, May 20. www.nationmedia.com/dailynation/nmgcontententry.asp? category_id=39&newsid=124032 (accessed May 25, 2008).

Loyn, D. 2007. "Good Journalism or Peace Journalism." *Conflict & Communication Online* 6(2).

Lynch, J. 2007. "Peace journalism and Its Discontents." *Conflict & Communication Online* 6(2).

Lynch, J. and A. McGoldrick. 2005. *Peace journalism*. Stroud: Hawthorn Press.

Miller, T. 2007. *Cultural Citizenship: Cosmopolitanism, Consumerism, and Television in a Neoliberal Age*. Philadelphia: Temple University Press.

Nohrstedt, S. A. and R. Ottosen. 2001. *Journalism and the New World Order: Gulf War, National News Discourse and Globalization*. Gothenburg: Nordicom.

———. 2007. War Journalism in Threat-Society: Peace Journalism as a Strategy to Challenge the Mediated Culture of Fear? Paper presented at the 20th Annual Conference of the German Peace Psychology Association, "Conflict, Communication and Intergroup Relations," June 15–17, 2007. University of Konstanz, Germany.

Omaar, R. and A. de Waal. 1993. "Somalia: Adding 'Humanitarian Intervention' to the U.S. Arsenal." *Covert Action* (44).

Opondo, O. and O. Orlale. 2008. "Bill to Reign In Errant Phone Users." *Daily Nation*, July 3, Politics section. Nairobi: Nation Media Group.

Orgeret, K. A. 2006. Moments of Nationhood: The SABC News in English—The First Decade of Democracy. Series of Dissertations Submitted to the Faculty of Humanities, University of Oslo Nr. 267.

Østerud, Ø. 2007. "Mild Imperialisme? Kan Vestlige Land Sivilisere Verden I Sitt Bilde? (Mild Imperialism? Can Western Countries Civilize the World in Its Picture)," *Klassekampen*, July 21.

Ottosen, R. 1995. "Enemy Images and the Journalistic Process." *Journal of Peace Research* 32(1): 97–112.

Ottosen, R. 2007. "Emphasis Images in Peace Journalism: Theory and Practice in the Case of Norway's Biggest Newspaper." *Conflict & Communication Online* 6(1).

Robertson, R. 1992. *Globalization: Social Theory and Global Culture*. London: Sage.

Rønning, H. 2005. "African Journalism and the Struggle for Democratic Media." In *Making Journalists*, ed. Hugo de Burgh, 157–181. London and New York: Routledge.

Sciades, G., ed. 2005. *From the Digital Divide to Digital Opportunities: Measuring Infostates for Development*. Paris: ITU/Orbicom-UNESCO. http://www.itu.int/ITU-D/ict/publications/dd/summary.html.

Seib, P. 2004. *Beyond the Front Lines: How the News Media Cover a World Shaped by War.* New York and Basingstoke: Palgrave Macmillan.

Taylor, P. 2007. *An Introduction to PSYOPS.* Leeds: Institute of Communication Studies, University of Leeds.

Teer-Tomaselli, R., H. Wasserman, and A. S. de Beer. 2007. "South Africa as Regional Media Power." In *Media on the Move. Global Flow and Contra-Flow*, ed. D. K. Thussu, 153–164. Milton Park: Routledge.

Thuku, W. 2008. Hate SMS suspect charged. *The Standard*, May 22, 2008. http://www.eastandard.net/InsidePage.php?mnu=details&id=1143987054&catid=4

Thussu, D. K. 2006. *International Communication: Continuity and Change*, 2nd edn. London: Arnold.

———. 2007. *Media on the Move. Global Flow and Contra-flow.* Milton Park: Routledge.

Washington, M. A. 2003. "Covering Global War on Terrorism through African and American Lenses." *The Perspective*, April 24.

Index

CPSIA information can be obtained
at www.ICGtesting.com
Printed in the USA
LVHW081945030522
717812LV00006B/451

9 780230 614864